3D Printer Projects for Makerspaces

3D Printer Projects
for Makerspaces

Lydia Sloan Cline

New York   Chicago   San Francisco   Athens   London   Madrid
Mexico City   Milan   New Delhi   Singapore   Sydney   Toronto

Sponsoring Editor
  Michael McCabe

Editorial Supervisor
  Stephen M. Smith

Production Supervisor
  Lynn M. Messina

Acquisitions Coordinator
  Lauren Rogers

Project Manager
  Patricia Wallenburg, TypeWriting

Copy Editor
  James Madru

Proofreader
  Claire Splan

Indexer
  Claire Splan

Art Director, Cover
  Jeff Weeks

Composition
  TypeWriting

To everyone who makes this world better
with their inventions and efforts.

About the Author

Lydia Sloan Cline teaches drafting, digital modeling, and 3D printing classes at Johnson County Community College in Overland Park, Kansas. She works for architecture firms, judges competive technology events, and is active in her local maker community. Lydia is also the author of *3D Printing and CNC Fabrication with SketchUp* and *3D Printing with Autodesk 123D, Tinkercad, and MakerBot* (published by McGraw-Hill Education).

Contents

Preface

3D PRINTING, A PROCESS THAT CREATES a physical model from a digital one, is increasingly popular with consumers, hobbyists, entrepreneurs, and businesses of all sizes. According to a 2014 Wohlers report, the global 3D printing market was valued at $2.3 billion in 2013 and is predicted to be $8.6 billion by 2020. Aerospace, automotive, medical, architectural, and high-tech industries routinely incorporate it in their design and manufacturing processes.

The Maker Movement Is Accessible to All

The beauty of the so-called "maker movement" is that the tools of manufacturing are not restricted to large companies anymore. Powerful, free software and relatively cheap tools are available to everyone (Figure P-1). Printers are sold at big-box stores and on Amazon. There are 3D printers at libraries, community makerspaces, and FedEx locations. Even the U.S. Postal Service is considering how to incorporate 3D printing. This accessibility, paired with funding sources like Kickstarter and GoFundMe, has enabled the launch of many successful small businesses.

What Can I Do with a 3D Printer?

You can use a 3D printer for decorative, useful, practical, iterative, and prototyping purposes. It's a tool that facilitates design, and helps you think of ways to convert ideas and raw materials into products and services. Find your "blue

Figure P-1 A child soars with her butterfly bike at the Kansas City Maker Faire.

ocean" and create new markets. To this end, things you can 3D-print include

- Replacement parts that are expensive or hard to find

- Novelty parts to work with existing systems, such as LEGO-compatible bricks

- Solutions to around-the-house or -office annoyances, such as the part a teacher designed to keep her computer mouse from falling off a keyboard tray or the harmonica holder a musician designed so that he could move around the stage

- Downloaded models that others have made

- Iterations, such as the plastic prototypes a jeweler makes before committing to expensive metals and the time casting them
- Marketing and presentation materials, such as the miniature firefighting wall systems one business makes before it builds full-size ones
- Small-run manufacturing items

Solving Societal Problems

3D printers are in many K–12 and college classrooms because the future of learning is the future of making things. Students are challenged to solve everyday problems with them (Figure P-2) and learn how to design items they can't just go out and buy. Making is not just about what we know; it's about what we can do with what we know. Larger social and environmental challenges such as income inequality, climate change, recyclability, and sustainability may be addressed. For example:

- The need for injection molding and the resources that it consumes may diminish.
- Large Delta printers (Figure P-3) may give citizens in developing countries sturdier homes printed from local soil.
- Startups are making renewable and recyclable filament out of food waste by-products and from hemp, the latter of which provides farming jobs as a side benefit. Other startups have developed machines that recycle plastic and make homemade filament, in an effort to make money and make the world a little better.
- Hyperlocal manufacturing is enabled, eliminating the need to ship products thousands of miles and saving resources consumed in the process.

Figure P-2   Entries by high school students for a 3D printing competition held at Johnson County Community College, Overland Park, Kansas. Models included spirograph wheels, phone amplifiers, and toys.

Figure P-3   A Delta printer that will make houses. (*Courtesy of wasproject.it.*)

That Competitive Edge

Human resources personnel speak of a growing need for all employees to understand the design-build process: what can and cannot be fabricated and how to handle production issues that require redesign work. Some companies even train their sales and marketing employees in 3D modeling and printing. At the very least, the discipline in learning the design and printing process can create better problem solvers and logical thinkers. The ability to design practical, printable parts is what makes 3D printers truly useful to the average person and is also what makes us technology *creators* instead of

just *consumers*. But since you're reading this, you already know that and are interested in participating in this new industrial revolution!

What's in This Book?

This book contains 20 tutorial projects. Each starts with a problem to solve, such as a personal, business, or education-related one. The solution is arrived at with different digital modeling programs and then printed. Settings and build plate orientation are shown. Some software is free. Other software has a cost but has limited-time free trial versions or is free/low cost to students and small businesses. Such software also may be available at your local makerspace.

Projects range from simple to complicated. The modeling process for all of them is covered in a tutorial, step-by-step manner. You may learn some new tricks in programs that you're already familiar with and get insights into programs that you've never used. Each project illustrates a slightly different technique or trick. We'll model and print all of them and post-process some. We'll do file conversions at online-convert.com, at 3dp.rocks, and with Inkscape. While most projects are modeled from scratch, we'll also download items from online repositories and edit them. When things don't go our way, either with modeling or printing, we'll solve those problems.

Each project is self-contained; you don't have to read prior projects to do it. Just browse for projects that interest you and jump right in! Each project has a "Things You'll Need" list, and vendor sources are listed in "General Stuff Before We Start." Some projects (and many others) are shown on my YouTube channel. Check it out at youtube.com/profdrafting.

What Software and Hardware Will We Use?

The projects will use software and hardware that you can reasonably expect to find in your home, school, or local makerspace. The modeling programs are Autodesk Fusion 360, Meshmixer, Tinkercad, 123D Design, AutoCAD, SketchUp Make and Pro, Inkscape, and Fuel 3D Studio. Please be aware that all these programs are constantly updated. Hence, by the time you read this book, there may be some changes to them already. Consider all the updates part of the challenge of being an early 3D printing technology adopter.

We'll use four different fused deposition modeling (FDM) printers:

- MakerBot Replicator 2 with a single extruder and cold acrylic build plate (Figure P-4)

- MakerBot Mini+ with a single extruder and cold build plate (Figure P-5)

- Gcreate Gmax 1.5XT+ with a dual extruder, heated aluminum build plate, and BL Touch autoleveler (Figure P-6)

- Lulzbot Taz 6 with a single extruder, heated borosilicate glass build plate, and sensor probes autoleveler (Figure P-7)

We'll use a Fuel 3D camera for a scanning project and silicone rubber for a casting project. We'll design for single- and multiple-color printing, and print in single and multiple colors (including multiple colors on a single extruder). We'll make prints that stand alone and prints that attach to objects, figuring out clearances in the process. Our filaments will be PLA, ABS, and bronze. We'll use three slicers, which are the programs that convert the digital model into language the printer reads. They are MakerBot Desktop, Cura, and Simplify3D. We'll also show slicer settings and file orientation for each project.

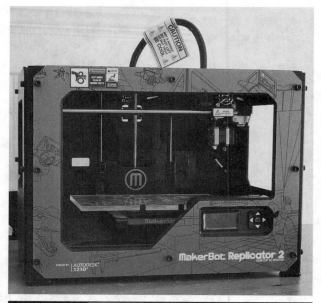

Figure P-4   MakerBot Replicator 2.

Figure P-5   MakerBot Mini+.

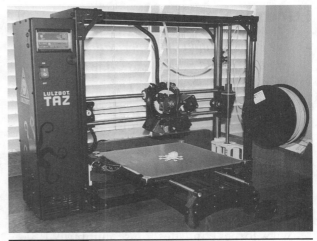

Figure P-7   Lulzbot Taz 6.

Where Can I Find Makerspaces?

If you don't have the software and equipment just described, you can still participate. Makerspaces, also called *hacker spaces*, *fab labs*, and *tech shops*, are community workshops with computers and fabrication tools that you can access free or for a monthly subscription. Some are privately owned; others are maintained inside public libraries and museums. Many makerspaces offer classes. Some are listed in the "Further Reading" at the end of this Preface; you can also Google "makerspace" or "hacker space" and the name of your city to find what's available near you.

Now join me in "General Stuff Before We Start," where we'll discuss filaments, build plates, and other odds and ends as a prelude to our projects.

Figure P-6   Gcreate Gmax 1.5XT+.

Further Reading

- *3D Printing and CNC Fabrication with SketchUp*, by Lydia Sloan Cline (TAB/McGraw-Hill, New York).

- *3D Printing with Autodesk 123D, Tinkercad, and MakerBot*, by Lydia Sloan Cline (TAB/McGraw-Hill, New York).

- Information on recycling plastic: http://preciousplastic.com/en/.

- 3D printing market forecasts: www.forbes.com/sites/louiscolumbus/2015/03/31/2015-roundup-of-3d-printing-market-forecasts-and-estimates/#2dcc808a1dc6.

- Giant Delta printer: www.3ders.org/articles/20140731-wasp-team-to-3d-print-homes-in-developing-countries-using-clay-and-soil.html.

- Services that match people with files and people with equipment: www.100kgarages.com/ and 3dhubs.com.

- Chart of design software: https://i.materialise.com/3d-design-tools.

- Modeling software comparisons: https://i.materialise.com/3d-design-tools#b-3d-design-apps and www.matterhackers.com/articles/finding-the-right-3d-modeling-software-for-you.

- Simplify3D slicer: simplify3d.com/buy-now/.

- MakerBot Desktop slicer: makerbot.com/desktop.

- Cura slicer: ultimaker.com/en/products/cura-software.

- Facebook groups: 3D Printing Club and 3D Printing.

- Popular modeling programs: SketchUp (sketchup.com), Blender 3D (blender.org), Tinkercad (tinkercad.com), Fusion 360 (autodesk.com), Solidworks (mcad.com), AutoCAD (autodesk.com), Inventor (autodesk.com), 3DS Max (autodesk.com), Maya (autodesk.com), Zbrush (pixologic.com), Sculptris (pixologic.com), Form Z (formz.com), FreeCAD (freecadweb.org), Rhinoceros 3D (rhino3D.com), and Microstation (bentley.com).

3D Printer Projects
for Makerspaces

General Stuff
Before We Start

WHILE A MAKERSPACE may have almost everything you need, it's likely you'll still have to set some things up or make some decisions and purchases yourself. So before starting our projects, let's cover some general subjects that apply to many of the projects in this book. In this way, you can return here to quickly find this information. We'll also cover some information on using SketchUp that is common to all our SketchUp projects.

Cartesian Printers

A *Cartesian printer* is named after the mathematical x, y, and z coordinate system. Those coordinates tell the printer's moving parts where and how to move. Cartesian printers have a square or rectangular build plate (Figure GS-1), and the plate's sides are the x and y axes. The plate moves as the print develops.

Figure GS-1 A Cartesian printer has a square or rectangular build plate.

Cartesian printers are either CoreXY or RepRap style. A CoreXY printer is more complex. It has pulleys and rods and long belts. Its build plate descends as the print is built, and the extruder moves along the *x* and *y* axes. A RepRap style has a single belt and a pulley. Its build plate moves back and forth, and its extruder just moves along the y axis. Generally speaking, a CoreXY printer is more beginner friendly, and doesn't take a lot of experimentation to produce high-quality prints. Examples of CoreXY printers are MakerBot and Ultimaker. A RepRap design printer may need more experimentation to get the same quality prints that a CoreXY produces right away. RepRap designs are popular for those seeking cheaper printer solutions, for example, greater build volume for the price. The Gcreate Gmax and Lulzbot Taz are examples of RepRap style.

Delta Printers

A Delta printer works within the Cartesian system but typically has a circular build plate that doesn't move (Figure GS-2). A circular build plate offers less print area for square parts than Cartesian printers do, but its advantage is that it works well for tall, thin prints. The extruder is suspended above the bed by three arms in a triangular configuration—hence the name *Delta*. These printers are designed for speed, so they typically have a Bowden extruder (discussed shortly).

Other Printer Features

There are many printer brands, and all have different features. One is the build plate: size, cold vs. heated, manually leveled vs. autoleveled. Printers have different construction types, extruders, and methods to hold the filament spool and thread the filament into the extruder. Some lack slots for SD cards; some lack USB

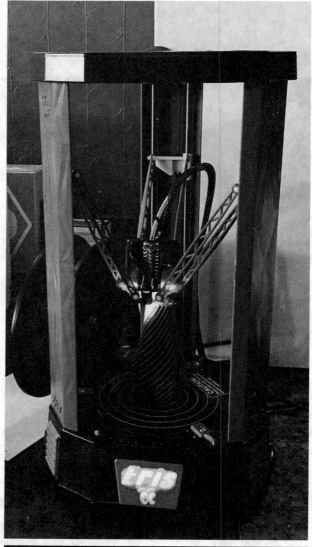

Figure GS-2 A Delta printer has three arms and a stationary build plate.

ports. Kits are popular because they are cheaper than fully assembled printers and assembling them leads to greater learning about how they work. You can also buy clones of well-known printer brands and even of some kits. Comparison websites are listed at the end of this chapter.

Once you work with multiple printers, you'll learn the strengths and weaknesses of each type. For example, one printer may tend to grind filament. Another may print small, sharp corners well while another prints them poorly.

Closed and Open Source

Printers may be closed or open source. Closed-source printers are either slightly modifiable or not modifiable at all (at least without voiding a warranty) and typically start as Kickstarter projects. They often require use of proprietary filament, slicer and firmware. Their drivers, slicers, and firmware upgrades may be delivered as push notices and automatically installed. Closed-source printers are usually highly developed, their software works well for their specific hardware, and technical support is available.

Most printers are open source and differ in their level of development and customer service. Some are perpetual works in progress. To update their drivers and firmware, you must go to third-party sites and download/install them yourself. Technical support often takes place in voluntary member support forums. Open-source printers tend to be RepRap style.

Firmware

All printers have *firmware*, which is software on the motherboard that makes the hardware work. Firmware interprets commands from the g-code file and controls the machine accordingly. It affects print quality; for instance, even though you can choose high, medium, and low print quality through the slicer, firmware affects a print's resolution. Developers update firmware just like they do slicing programs. Marlin, Sailfish, Repetier, and Smoothie are popular firmware programs.

Extruders

All printers have an extruder, also called a *print head*, which is the spout through which the melted plastic flows. Extruders are either all metal, which can handle the high temperatures some filaments require, or metal and plastic, which print at lower temperatures. The *hot end* is the active part that melts the filament and pushes it through the nozzle. It consists of a nozzle, heating cartridge, and heating barrel (Figure GS-3). The *cold end* is the rest of the extruder.

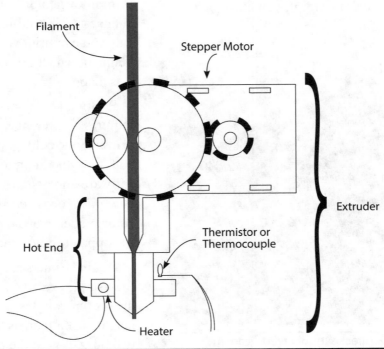

Figure GS-3   Extruder diagram.

Extruders are *direct* or *Bowden*. Direct extruders feed the filament straight into the hot end (Figure GS-4). This enables the extruder to retract, or pull the filament back, more quickly when crossing open spaces. A Bowden extruder physically separates the hot and cold ends. The cold end is mounted behind or inside the printer (Figure GS-5), and the filament travels through a tube to the hot end. This makes the hot end lighter, and hence it moves faster, lessening the printing time. A Bowden's disadvantage is that flexible filament can't be used in it and switching filament midprint is impossible.

There are many styles of direct extruders, and they're made for specific machines. Common ones are the MK8 (for the MakerBot), MK10 (for the Flashforge, Dremel, and Wanhao), and RepRap (for many open-source printers). There are different hot-end styles, too. The E3D is a popular all-metal hot end, and the Diamond is a hot end that can print three separate colors from one nozzle.

There are single and dual extruders (Figure GS-6). A single extruder prints in one color. To obtain multiple colors with it, you must pause

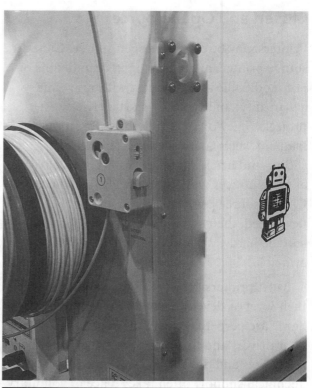

Figure GS-5 A Bowden extruder physically separates the hot and cold ends.

the print, unload the filament, load the new color, and restart the print. A dual extruder can print in two colors without midprint filament changes. Using a dual extruder is also the only way you can print with a color and dissolvable filament, which is how the print in Figure GS-7 was made.

Printing two colors with a dual extruder requires a different design approach than printing one color. Typically, colors are separate STL files that are manually aligned in the slicing software and assigned different extruders. An *ooze shield* or *prime pillar* is printed with a two-color model to avoid each color dribbling onto the opposite color parts.

Dual extruders are heavy, so they run slower than single extruders. Their larger size also means that you lose some of the build area and height. Then there is the challenge of aligning two nozzles with the build plate. They must

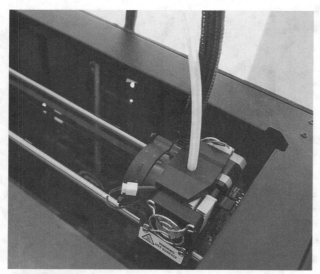

Figure GS-4 A direct extruder feeds filament straight into the hot end.

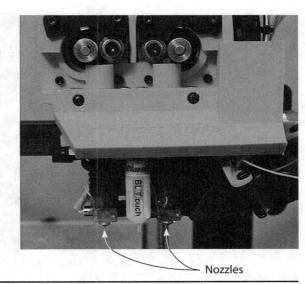
Nozzle

Nozzles

Figure GS-6 A MakerBot Replicator 2 single extruder (*left*) and a Gmax 1.5XT+ dual extruder (*right*).

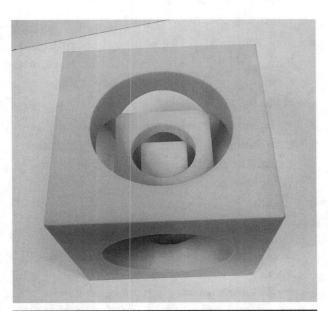

Figure GS-7 This print was made with dissolvable supports.

be aligned even if you're only using one color, because an unaligned nozzle might knock the print over during construction.

Nozzle Size and Layer Height

The *nozzle* is the part of the extruder through which the melted filament emerges. Nozzle

openings range from 0.1 to 1 mm in diameter, with 0.4 mm being the most common. Smaller nozzles enable more detail but take longer to print. Large nozzles don't enable fine detail but print faster. Support material breaks off easier when made with small nozzles.

Nozzle size is irrelevant to filament size but is relevant to *layer height*, which is the thickness of the filament string deposited. Set an appropriate layer height for the nozzle size. For instance, you wouldn't want to pair a small layer height with a large nozzle or a large layer height with a small nozzle. A 0.8-mm nozzle and 0.6-mm layer height make a fast, rough print. A 0.4-mm nozzle and 0.2-mm layer height make a slow, finely detailed print. Models with lots of overhangs print better with lower layer heights. Models with bridges print better with higher layer heights.

Standard nozzles are made of brass. Metallic filaments wear the nozzle's opening out faster, so consider a nickel-plated brass or hardened-steel nozzle (Figure GS-8) when printing with those filaments. Like extruders, nozzles are made for specific printers, so make sure that you buy an appropriate one.

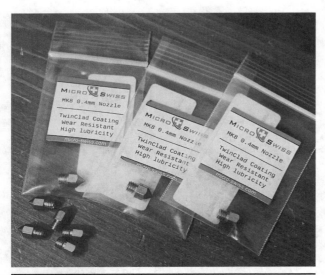

Figure GS-8   Brass and nickel-plated brass nozzles for MakerBot printers.

Build Plates

The build plate, also called the *printer bed*, is the horizontal surface where the print is made. There are cold and hot plates. Cold plates are room temperature; hot plates are the ones for which you can set a temperature. While any filament can be printed on a hot plate, hot plates are specifically needed for some filaments to prevent warping during the printing process. For example, ABS filament is 200°C when extruded out of the nozzle, but quickly warps once it hits the build plate's cooler surface. A hot plate helps to prevent warping because it keeps the lower layers of the print warm as the hotter top layers are extruded on top of them, allowing a more even overall cooling.

All build plates must be leveled, that is, positioned horizontally so that all points on it are the same distance (a business card thickness) from the nozzle. Some plates are leveled manually. Others are autoleveled, meaning that a probe or sensor polls four points on the bed to see how the corners compare, and then the printer's firmware adjusts for discrepancies with small z-axis movements as the extruder moves around. *Babystepping* refers to fine-tuning the extruder's position from the plate via the LCD (liquid crystal display) screen while printing the first layer. Both automatic- and manual-leveling build plates have screws underneath them that you turn to set the plate's distance from the extruder.

The most common build plate materials are

- *Acrylic* (for cold printing). This is inexpensive and may warp with use. Sand with a fine-grit paper to improve print adhesion.

- *Plain glass* (for cold and hot printing). This is inexpensive and can break easily when removing a stubborn print.

- *Borosilicate glass* (for cold and hot printing). This has a high heat tolerance and resistance to temperature changes and hence less warping and breaking.

- *Aluminum* (for cold and hot printing). This distributes heat more evenly than glass and doesn't break. It can warp but is easier to straighten out than glass.

Some makers place glass on top of aluminum to combine aluminum's heat distribution and glass's tendency to remain straighter over time. Some even sandwich cork in between for insulation and hold it all together with binder clips, one placed on each side of the plate. When using clips, ensure that they don't interfere with the extruder's motions because the extruder goes beyond where the print is located. Watch the printer in operation to see where the extruder moves, and then select a spot for each clip.

Adhesion Covers and Treatments

You can print directly on the build plate or apply papers, films, and adhesives to help the print stick. Some treatments only work on cold plates, some only work on hot plates, and some work on both. Filaments brands, types, and even colors adhere differently to different plates and surfaces, so experiment. An appropriate combination of plate and surface treatment avoids the need for a raft (discussed later) to adhere the print to the plate.

> **Tip:** Sometimes prints come off easily, and sometimes they don't. Wait until a heated plate has completely cooled before removing the print. Try putting a stubborn print in the freezer for a few hours and remove it after the cold temperature has caused the plate to contract away from the print.

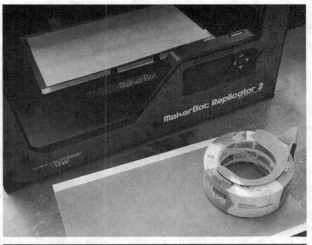

Figure GS-9 A sheet of painter's tape on an acrylic build plate. Painter's tape comes in sheet and strips.

Figure GS-10 A sheet of BuildTak on an acrylic build plate. It comes in multiple sizes.

The following are popular treatments:

- *Blue painter's tape.* This is adhesive-backed paper that comes in multiple sizes and strip widths. It is used on cold plates (Figure GS-9).

- *BuildTak sheet.* This is adhesive-backed plastic that comes in multiple sizes (Figure GS-10). It's more durable than painter's tape and is used on both cold and hot plates. It's also difficult to remove.

- *PEI sheet.* This is thin, adhesive-backed plastic that is cut to size and used on both cold and hot plates.

- *PRINTinZ skin.* This is a thin, adhesive-backed, glass-reinforced sheet used on both cold and hot plates.

- *PRINTinZ ninja plate.* This is a thick, glass-reinforced sheet that sits on top of the build plate and is removed and flexed when the print is finished. It is used on both cold and hot plates.

- *Elmer's purple glue stick.* Rub this adhesive onto a bare plate or onto painter's tape. It is used on both cold and hot plates.

- *Uhu glue stick.* Rub this adhesive onto a bare plate or onto painter's tape. It is used on both cold and hot plates.

- *Aqua Net unscented hairspray.* Lightly spritz this onto a cloth and wipe onto a bare acrylic or glass plate or over painter's tape. Let it dry before printing. It is used on both cold and hot plates.

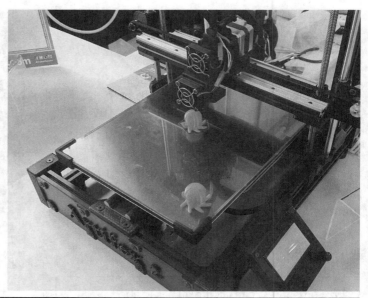

- *Kapton tape.* This is an adhesive-backed polyimide (plastic) that comes in thin and wide rolls (Figure GS-11). It's used on heated plates.
- *ABS slurry.* This a homemade solution of ABS scraps dissolved in acetone and wiped onto a bare build plate or onto Kapton tape and then let dry. Don't dissolve too many scraps because a thick solution may make prints impossible to remove. It is used on heated plates.
- *Elmer's glue slurry.* This is a homemade solution of one part Elmer's glue dissolved in ten parts water. Wipe onto a bare build plate or onto Kapton tape and let dry. It is used on heated plates.

Here are some additional notes about covers and treatments:

- All build surfaces must be clean; touching them leaves dirt and oils to which filament can't stick. Clean before each print, and remove any old residue. Wipe bare and covered plates with an ammonia-based glass cleaner or isopropyl alcohol. These come in spray bottles and wipe packets and work by degreasing the plate and then evaporating. Both treatments provide squeaky-clean plates, which alone can provide sufficient adherence.
- Tall, thin prints with little surface contact area need more adherence than just a clean plate provides. Use hairspray or glue for them.
- Tape comes in sheets and strips. Large sheets cover the whole plate at once but can be hard to apply without wrinkles and bubbles. Strips are easier to apply. When using strips, tiny gaps between the strips are fine, but overlays are not because they create an uneven printing surface. Rubbing painter's tape with isopropyl alcohol slightly dissolves the top, exposing a gummy residue, and that may provide sufficient adherence.
- The specific brands and products mentioned, such as Aqua Net unscented and Elmer's purple glue stick, have been proven to work for our purposes. Many brands or other products by those brands do not.

Tip: All adhesive-backed papers get worn or damaged with use. Place inexpensive painter's tape over expensive surfaces such as PEI and BuildTak when cold plate printing to preserve those expensive surfaces.

Filament

There are many brands and types of filament, and more are brought to market each day (Figure GS-12). Choices include colors, transparent, glow-in-the-dark, high-strength, water-soluble, wood, metal, glass, nylon, stone, ceramic, flexible conductive, temperature color changing, and sustainable.

Not all filaments work in all printers; some printers accommodate many filaments, and others accommodate only one or two. Check the filament manufacturer's website for a list of printers in which their product has been successfully tested and for recommended temperature and speed settings. Filament manufacturers give temperature ranges because the ambient room temperature, printer model, nozzle size, speed settings, and layer height all make a difference. Some filaments work better in different size nozzles, and colors within brands may print optimally at different temperatures (e.g., the lighter the color, the less heat is needed). Also check size; filament is either 1.75 or 3 mm in diameter, and all printers use either one or the other.

Filament Storage

Exposure to moisture causes filament to swell, which jams the extruder. Exposure to ultraviolet (UV) light makes it fragile. Store filament in airtight bags and out of sunlight. Rubber storage bins or lidded 5-gallon paint buckets also work well. Throw some desiccant packets in with the filament. Some packets have silica beads that change colors as they absorb moisture. Filament should be used within a year of its purchase because it degrades with time.

Tip: A clue to water-saturated filament is the hissing and popping sounds made as the water heats up inside the extruder and becomes steam. Before throwing it out, try putting it in the oven for 3 hours at 60°C to dehydrate.

Popular Filaments

Polylactic Acid (PLA)

This is corn based and biodegradable, and it emits a small amount of fumes. It doesn't require an enclosure or a ventilated room and can be printed on a cold or hot build plate. PLA prints at temperatures between 190 and 230°C and is brittle, hard, and shiny. It's a good beginner filament, suitable for trinkets and prototypes. Small items (under ~170 mm long) print well on a clean, uncoated glass or acrylic plate. Hot plates offer better adhesion for larger prints; set the plate's temperature between 40 and 50°C.

Figure GS-12 | Filament from Inland, ColorFabb, MakerBot, and Hatchbox.

A PLA print degrades rapidly when exposed to moisture and sunlight, so don't put it in a dishwasher, pour hot liquids into it, or use it in areas that build up heat, such as inside a car.

PLA is food safe (excluding filaments with specific additives; check the label or the manufacturer's website), but bacteria builds up in a print's gaps and spaces. Antibacterial filament brands and food-safe sealant coatings exist. Printers that use PLA and ABS filament may also emit ultrafine particles that are unsafe for ingestion. Commercial Selective Laser Sintering (SLS) printers are more suitable for producing food-safe eating utensils. Any printer used for food products should have a lead-free nozzle (stainless steel is best), be dedicated to food products so that they won't have residue from non-food-safe filaments, and be in a clean environment.

Acrylonitrile Butadiene Styrene (ABS)

This is petroleum based, not biodegradable, and emits strong fumes. It needs to be printed in a well-ventilated room and preferably with an enclosure to maintain temperature and contain fumes. An ABS print can be smoothed in an acetone bath, which removes the layer lines, an advantage for artistic prints.

ABS is stronger and harder than PLA, makes better supports and bridges, and is flexible. It's suitable for functional parts and has a longer service life than PLA. However, it does degrade with continuous exposure to sunlight. It is not food safe because it has toxic chemicals that could leach into food. It is more difficult to work with than PLA due to its tendency to warp while printing. It also shrinks more than PLA, a consideration when making interlocking parts. Such parts typically need a 0.4- to 0.5-mm clearance. ABS parts warp or shrink differently depending on the shape and size of the piece,

so it may not the best choice for projects that require joining multiple pieces. In such cases, you can expect to post-process with gap fillers and sanding.

ABS prints best between 235 and 256°C, with a heated build plate temperature between 80 and 110°C.

Polyethylene Terephthalate (PET, PETG)

This is glossy, semitranslucent, strong, impact resistant, and flexible like ABS but without the warping and odor issues. It's also food safe. PETG prints on a heated plate in any printer that can print ABS, and it bridges well. However, it doesn't have the heat resistance of ABS, so it can't be exposed to hot sunlight all day. It prints at about 60 percent of the speed of ABS and doesn't solvent bond with common, safe solvents. It also can't be smoothed with an acetone bath. PET/G is for engineering-grade objects and functional models such as mechanical parts, robotics, drones, phone cases, wearable technology, and screw threads. It prints best between 220 and 250°C with a heated bed of 70 to 85°C.

Polyvinyl Alcohol (PVA)

This is used as a dissolvable support in dual-extrusion PLA printing. It's odorless, nontoxic, and dissolves in water. It's good for complex prints with odd angles and overhangs. It prints at temperatures between 190 and 220°C, with an optional heated bed at 40°C. PVA is printed with PLA because they have similar printing temperatures.

High-Impact Polystyrene (HIPS)

This is used as a dissolvable support in dual-extrusion ABS printing. While it can be broken off by hand, it dissolves in Limonene, a colorless solvent made from citrus fruit, and takes between 8 and 24 hours to do so. It can also be used as a primary material. HIPS is good for complex prints with odd angles and overhangs. It emits strong odors, so it needs to be printed in a well-ventilated room. It's non–food safe.

HIPS prints at temperatures between 230 and 250°C with a heated plate temperature of 115°C. Wait until the print is fully cooled before removing it from the plate. It is used with ABS and PETG because they have similar printing temperatures.

Thermoplastic Elastomer (TPE)

This is a flexible filament used for bendable items. But know that if the model is thick and has a high infill percentage, you'll still get a pretty rigid "flexible" print. TPE makes a poor support material, so don't use it on prints that need support.

Wood (PLA Infused with Particles and Binders)

This is wood fibers and binders in PLA. A high-quality filament looks and smells like wood. It can be sanded and varnished. Editing slicer settings for temperature changes during the printing process results in tones and colors that resemble wood grain because the hotter the temperature, the darker the layers become. It's good for natural-looking furniture and musical and decor pieces. Print it between 180 to 245°C on a cold plate.

Bronze (PLA Infused with Bronze Powder)

This is hard, durable, and good for jewelry, statues, home hardware, and artifact replicas. This filament looks dull after printing and requires extensive sanding and polishing for an actual bronze appearance. Print it between 195 and 220°C on a cold plate.

Changing Filament

Load filament at the start of a new project and unload it at the end of a project to store. Purging old filament works well when it's done in the raft of a new project. Before loading new filament, make a clean cut at the end to insert into the extruder.

You might change filament midproject to incorporate multiple colors or when the filament spool is almost empty and you have to load a new one to complete the print. Some machines are easier to do this on than others. The MakerBot has a Pause function in which the extruder moves aside, and a Change/Load/Unload function that makes changing filament easy (Figure GS-13). Other printers have a Pause function, but the extruder stops in place, and there's no Load/Unload function, so purging the existing filament to make way for the new one is very difficult. Know that you'll ruin the print if you inadvertently move the extruder during this process. Difficulty in changing filament mid-print is a particular drawback on printers designed to accommodate large prints, because spool changes are expected on such prints.

The design has to be appropriate for a filament change. For instance, in the armchair coaster shown in Figure GS-14, the cushion of the chair was completed, and then the filament was changed to complete the higher armrests.

Change filament quickly because the motor may time out if you wait too long, or the idle extruder may clog if the filament cools too high up inside it. Extruder jams are often caused by heat creep. This occurs when printing at very slow speeds or while the printer is sitting idle after a print or during a color change and the hot end is still warm. Heat creeps up the extruder and melts the filament higher up inside it than it should be melted, which creates a hardened blob that jams the extruder. To help avoid this, purge the filament after each print.

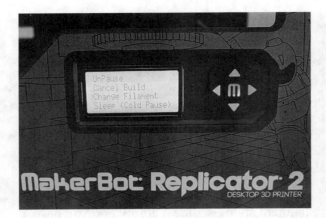

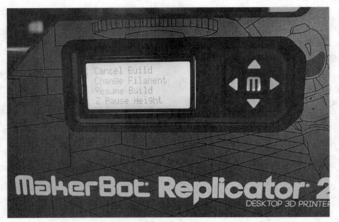

Figure GS-13   Accessing the Change Filament function in a MakerBot directly from the LCD screen.

Figure GS-14   The new color is layered on top of the
old color.

Glue and Solvent Welding

Models too large for the build plate can be printed in pieces and then glued together. The gel version of plastic epoxy or cyanoacrylate (Super or Gorilla Glue) and polystyrene cement (industrial plastic adhesive such as 3M Scotch-Weld) work well on PLA and ABS. Flexible and nylon filaments don't glue together well. Taulman, a manufacturer of nylon filament, makes a special glue for nylon products.

Most filament can be solvent welded, which fuses the pieces together by temporarily softening and dissolving the plastic, creating a stronger bond than glue. Loctite 401, Tenax 7R, and Plastruct Plastic Weld are popular solvent-welding products. PLA and ABS can be solvent welded; PETG cannot. A 3D printing pen loaded with the same filament the parts are made of can also weld them together.

After gluing, plastic filler such as Bondo is useful for hiding seams and filling in holes. Apply it in layers, and sand after it dries.

Slicer Settings: Dialing It In

The slicer is the software program that generates supports, has setting options for the physical print, and turns the model into g-code, the language that 3D printers read. Proprietary slicers can only be used with one printer brand, or can't be altered. Open-source slicers can be used with multiple brand printers and have communities that contribute to new releases. You may find that certain slicers work best with certain printers, may produce a better print on a specific file, or may generate better supports.

Most slicer programs are free. Popular ones include MakerBot Desktop, MakerBot Print (for the Plus generation of printers) Slic3r, Cura, MatterControl, Skeinforge, Craftware, and Simplify3D (S3D). All have different interfaces, options, and capabilities. S3D is a pay program that lets you do things others don't, such as stop the print at a specific layer to change filament or delete the otherwise undeletable raft the MakerBot Mini builds on all prints. S3D also has a dual-extrusion wizard that makes slicing two-color prints easy. Cura has interesting experimental settings, such as Wireframe. MakerBot Desktop is the most user friendly to a beginner.

Experiment with slicer settings for each printer, filament type, and even filament brand you use; the same filament type across brands may perform optimally at different settings. Each project in this book includes settings, but they may not be optimal with your printer and filament. Once you find a filament brand you like, sticking with it reduces the need for constant experimentation. Be aware that many brands are counterfeited, so it's best to buy directly from the manufacturer or a trusted source.

Here are settings you can find in any slicer:

- *Temperature.* This is the amount of heat applied to the filament. A too-low temperature will cause the filament to jam in the extruder or the print to warp on the build plate (if it's a warp-prone filament). A too-hot temperature will make the filament stringy and thin. PLA's recommended temperature range is 215 to 235°C. It can print with or without a heated build plate. If you choose to use a heated plate, its temperature range is 50 to 60°C. ABS's recommended temperature range is 230 to 240°C with a heated build plate at between 100 and 110°C. A heated build plate is set to a lower temperature than the printer.

- *Speed.* This is the rate at which the molten filament flows through the nozzle and is typically between 50 and 150 mm/s (meaning 50 to 150 mm of filament comes out every second). Some filaments need to be run through the extruder at slow speeds; others at faster speeds. Slow speeds create higher-quality prints; fast speeds create coarser ones. A common speed is 50 mm/s.

- *Extrusion.* This is the amount of plastic that comes through the nozzle. If it's stringy or oozing too much, make small adjustments to the multiplier setting. The common multiplier default for PLA is 0.9 and for ABS is 1. Increasing it from 1.0 to 1.05 extrudes 5 percent more filament. If you increase the multiplier, you might need to increase the temperature, too.

- *Retraction distance.* This is how much molten filament the extruder motor sucks back up into the nozzle during nonprint moves to avoid plastic drips on open spaces and cobwebby strings between parts. If you see blobs and strings, increase the retraction setting (or turn it on if it's turned off). Bowden extruders usually need a higher retraction distance than direct-drive extruders do. Increase this by 0.5- or 1-mm increments at a time.

- *Retraction speed.* This is how fast the filament is pulled back from the nozzle. A common retraction speed range is 1,800 to 6,000 mm/min or 30 to 100 mm/s.

- *Coasting.* This is when the extruder stops printing a specific distance before a nonprint move, letting leftover filament clear before retraction starts. A common coasting default is between 0.2 and 0.5 mm. A coasting distance of 5 mm means that the nozzle will not extrude filament for the last 5 mm before the end of a perimeter, and leftover filament in the hot end is carried for the last 5 mm. If this setting is too high, the print will have gaps and underextruded areas.

- *Layer height.* Also called *resolution*, this specifies the height of each filament layer. Prints made with thin layers (~0.1 mm)

Figure GS-15   This pencil holder was made with a low layer height to show detail.

and small nozzles are more detailed (Figure GS-15) but take longer. Detail on designs with sloped surfaces is harder to get than on designs with flat surfaces. Prints made with thick layers (~0.2 mm) are coarser but print faster. They're useful for prototypes and prints without much detail.

- *Initial layer thickness.* Also called *first layer height*, this is the thickness of the first layer on the build plate. You can make it thicker than the rest to provide a sturdier base for the print. A common default is 0.3 mm. You can also adjust the speed of the first layer height. Making it slower, such as 30 to 50 percent of the regular speed, will help it to adhere to the build plate.

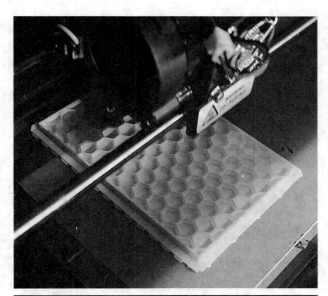

Figure GS-16 This is a 10 percent infill generated with the MakerBot Desktop slicer.

- *Shell.* This is the print's outer wall. The number of shells is the number of times the printer traces that wall before constructing the inside. Shell (i.e., wall) thickness can also be adjusted; increasing it makes it thicker and the print stronger. A common thickness default is 0.8 mm.

- *Infill.* This is the density of the model's solid parts. It is measured in percentage, not millimeters like the other settings are. A 0 percent infill is hollow; a 100 percent infill is completely solid. Everything in between generates a pattern (Figure GS-16). The higher the infill, the stronger and heavier the print will be, and the longer it will take to print.

- *Fan.* This setting applies to the fan that cools the print. Cooling is particularly needed on overhangs, bridges, and small, detailed prints. If details look muddy, it may be because one layer doesn't have time to cool before the next layer is applied. You can speed up, slow down, and even turn the fan off. Adjusting the fan setting is more effective on PLA prints than on ABS prints.

- *Supports.* These are structures that hold up *overhangs*, which are parts of the model with nothing underneath (Figure GS-17). Generally, features that are angled 45 degrees or more need supports; features angled less than this don't. Slicers generate supports automatically but let you finesse their placement and adjust their density. Supports can be difficult to remove and often leave small scars on the print.

- *Brims.* These are lines of filament around the bottom of a print that anchor it to the build plate. Sometimes they go under small parts of the model or under supports. Figure GS-17 shows brims at the bottom of the supports. The brims are resting on a larger raft.

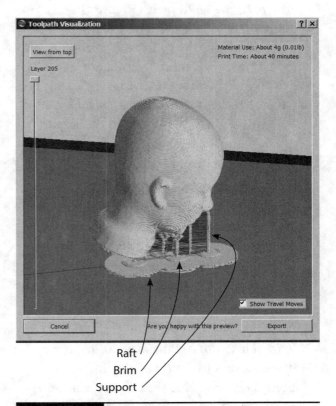

Raft
Brim
Support

Figure GS-17 This preview, generated in a MakerBot Desktop, shows supports and brims to be printed. Both are on top of a raft.

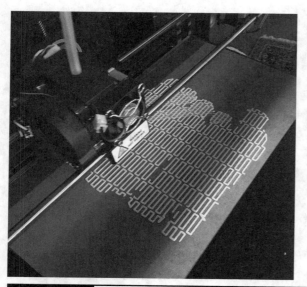

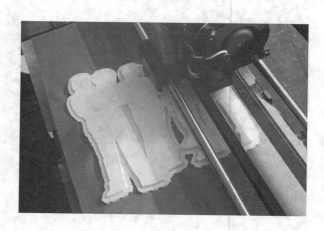

Figure GS-18   Raft under construction (*left*). Model on top of the raft (*right*).

- *Raft*. This is a thick lattice covered with a few layers of filament that helps the print to adhere to the build plate (Figure GS-18). A common default is 0.4 mm. Ideally, a raft will peel off the print easily; if it doesn't, try increasing the distance from the print's first layer. A Dremel tool is useful for sanding parts of the raft that stubbornly stick to the print. Don't use rafts on multiple-part prints that will be glued, because if the rafts can't be entirely removed, the parts won't fit together well.

- *Bridges*. These are horizontal overhangs. The extruder pulls filament from one end to the other to "bridge" the overhang (Figure GS-19). It can do this because the cooling fan makes the plastic quickly semirigid, and the extruder movement holds the bridge in tension until it anchors on the opposite side.

- *Skirt*. This is an outline around the model. Its purpose is to get rid of any filament clumps or dirt inside the extruder and start a smooth flow of filament. Some slicers just extrude a straight line before starting the print.

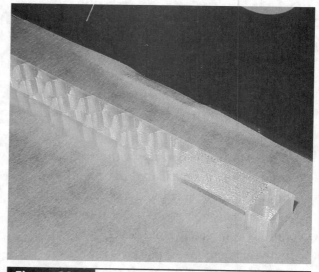

Figure GS-19   A bridge is a horizontal overhang that is spanned by filament.

Test your settings with calibration models downloaded from Thingiverse.com. They let you see how well both your printer and filament handle different shapes, corners, tips, and small details. The Benchy (benchmark) boat is a popular calibration model. Make incremental changes, record the results (Figure GS-20), and then save the settings with different slicer profiles. Change just one setting at a time to

Figure GS-20　These Benchy model settings were recorded for reference.

know for sure what helped. Print a calibration cube and measure it with a digital caliper to see if it is the size it's supposed to be (Figure GS-21).

If you consistently get bad prints, try leveling the build plate, tightening all belts and bolts, adjusting the settings, changing the slicer,

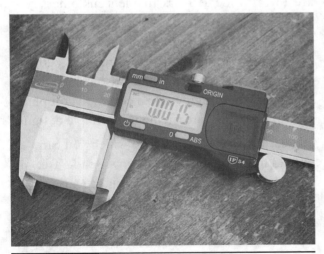

Figure GS-21　A cube whose digital measurements are 1″ on all sides should print that exact size.

and/or building an enclosure around an open printer. An enclosure helps to keep temperature constant and avoids drafts. It can be anything from an elaborate fabricated structure to a large cardboard box.

Printing Tethered versus SD (Memory) Card

Running the printer from a computer USB port is quicker when experimenting with slicer settings than copying each test print to an SD card. Some printers may also work more reliably when tethered. However, this requires a continuous data stream from computer to printer, and can be interrupted if the computer goes to sleep, does an automatic operating system update, has a power interruption, goes into power saver/screen saver mode, is inadvertently unplugged, or receives ambient interference (e.g., TV or radio transmissions, solar magnetic storms, lightning, or radar).

Tip: Different slicers have different capabilities in maintaining a data stream or in even just connecting the computer to the printer. In my experience with the three slicers used in this book, MakerBot Desktop maintains the most reliable stream. Simplify3D tends to drop the stream or have problems connecting. Pre-2.5 Cura versions have no built-in connection for USB at all. Know that you can use MakerBot Desktop to stream Simplify3D-sliced files to the printer. Export the S3D file as an X3G file, import the X3G file into MakerBot Desktop, and then send the file to the printer.

A print that takes more than 2 hours to complete is typically more successful when printed directly from an SD card (Figure GS-22). But be aware that prints made from SD cards can, and do, fail too, if the printer perceives a defect in the card during the printing process.

Running Multiple Printers Simultaneously

To run multiple printers simultaneously, it may be best to attach a computer to each one because running multiple printers from one computer drains processing power. That computer could be a Raspberry Pi (a tiny PC) or an inexpensive laptop. If you do run multiple printers from one computer, attach each printer to its own port.

Figure GS-22   An SD card inserts into a slot for direct printing.

Networked makers use a wifi connection and Octoprint, which is open-source slicing software that runs multiple printers with one Raspberry Pi. Octoprint also lets you monitor and control your printer remotely, which is handy for prints that take many hours to finish. Then there's OctoPi, which is a Raspberry Pi with Octoprint installed for greater simplicity.

Tip: Whether using a USB port or SD card, make sure that the slicer's settings include the specific printer used so that the printer can "home" properly. If you have the MakerBot Desktop slicer installed, turn off its background service (Services menu/Stop Background Service) to enable other slicers to work on the same computer.

Selling Your Work

Makers often explore how to make money with 3D printing, or at least how to make enough to pay for filament. One way is to create an online store at Shapeways (shapeways.com), a company that prints and ships the product. The maker gets a commission for each print sold. Some makers work with 3D Hubs (3dhubs.com), which are a worldwide network that lets customers get their own designs printed locally. Common Hubs requests include printing projects for students and prototypes for inventors. Makers who are good with design software can also charge customers for design work; a common complaint of Hubs makers is receiving unprintable designs.

There are other websites to which you can upload files for customers to download—free or for a price you choose—to print themselves. Cults3D (cults3d.com/), Pinshape (pinshape. com), My Mini Factory (myminifactory.com/), and CG Trader (cgtrader.com) are some. Yeggi (yeggi.com) and STL Finder (stlfinder.com), which are search engines for 3D printable models, will turn up more. Some sites have a streaming option; the customer streams the file directly to a personal printer or a service bureau instead of taking possession of the file. This is good for makers who don't want to release their files but is less popular with customers because they have no control over the slicing. Half of all transactions typically occur in the first two months after listing, so posting new content regularly keeps an income stream alive.

Learn what people want to buy. For example, vape mod boxes, cosplay items, and parts for racing drones, quad copters and Nerf guns are popular 3D printed items. The Pinshape website once blogged that its most popular items are toys, games, and home decor and their least popular items are jewelry and fashion. As with any sales venue, detailed descriptions and attractive photos grab viewer attention, which leads to more sales. Include photos of a printed model, not just computer renderings and print settings. Best-selling models tend to be priced for the impulse buyer.

Being Green

Many makers are ecoconscious and seek ways for their 3D printing to leave a smaller carbon footprint. Modeling the print well and disposing of scraps responsibly are the main ways to achieve this.

Modeling Well

A way to not waste plastic is to create digital models that can be printed successfully. Know what your printer is capable of; for example, the design should incorporate the level of detail the printer is capable of. Pay particular attention to

- *Wall thickness.* Walls must be thick enough to withstand removal from the build plate, removal of supports, any post-processing, and shipping. Supported walls should

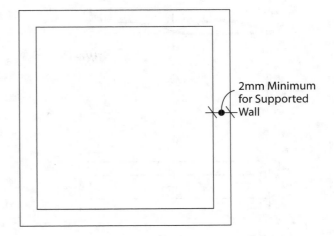

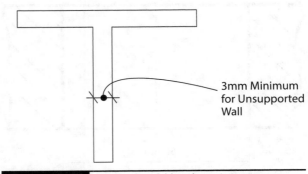

Figure GS-23 Walls should be the minimum printable thickness.

be at least 2 mm thick, and unsupported walls should be at least 3 mm thick (Figure GS-23). Thinner walls are likely to break, especially when supports are removed. Files that are scaled down to fit on a build plate may need to be revisited in the original modeling software for thickening.

- *Clearance.* This is space that allows fit between interlocking parts (Figure GS-24). Exact clearances vary with the printer, slicer, and filament, so experiment with small parts before printing large ones. On large models, cut out the interlocking parts, and just print them to get the clearances right before printing the whole thing. Generally, clearances are between 0.25 and 0.75 mm depending on the size and shape of the model. ABS typically needs between 0.4 and 0.5 mm. On

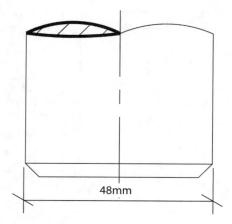

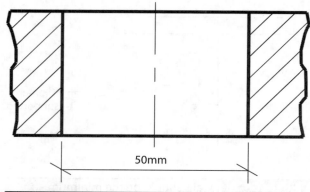

Figure GS-24 Clearance is the space that allows fit between interlocking parts.

small items, start with a clearance of 0.15 mm. Look for recommendations on the filament maker's website.

> *Allowance* is the amount of intentionally designed deviation between two interlocking pieces. *Clearance* is the space between two interlocking pieces. *Tolerance* is the amount a dimension may deviate from the design.

■ *Orientation and support.* Orient the print to avoid or minimize supports because supports may fail during the printing process or be difficult to remove. Simply flipping a model over, as shown in Figure GS-25, can greatly cut down the number of supports needed. Tree supports are often better than the vertical ones because they require less material. You can also try incorporating supports into the design.

■ *Orient for strength.* Parts that need to be strong should print laterally on the *z*-layer, not vertically across multiple layers. For example, if you printed a straw vertically, the

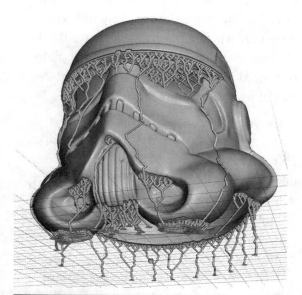
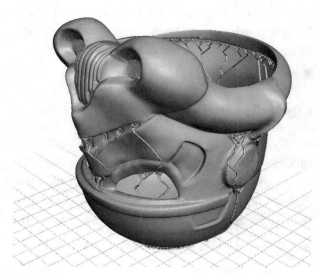

Figure GS-25 A helmet oriented straight up needs a lot of support (*left*). Turned upside down, it needs less (*right*). These tree supports were generated in Autodesk Meshmixer.

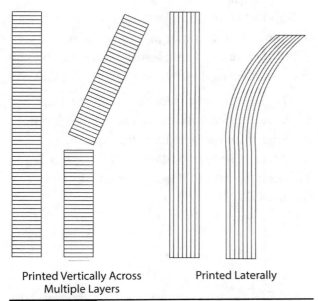

Printed Vertically Across
Multiple Layers   Printed Laterally

Figure GS-26   Print laterally for more strength.

straw would easily snap at each layer (Figure GS-26). Printed horizontally, the layers span the entire length, making it stronger.

- *Infill.* Thin, delicate prints need a high infill (75 to 100 percent) or they will break easily. But 100 percent infill is not good for thick prints because of heat buildup during the printing process. In such cases, holes must

be incorporated into the design to let hot air escape. Most thick prints are fine with a 10 to 15 percent infill. A working part that will be under stress may need a 40 percent infill and an extra shell.

- *Dematerialize.* Reduce your designs' overall size, weight, and number of materials and hollow parts out before printing, if appropriate.

- *Filament type and quality.* Appropriate filament makes a print successful. Should the print be rigid or bendable, used indoors or outdoors, have force applied to it, be post-processed? Choose a filament with the needed properties, and then experiment with settings because different brands may print at different temperatures. Some brands also work better than others due to different additives, powder, and pigments or have higher quality-control standards for diameter consistency, impurities, and spooling (Figure GS-27). Spot-check filament diameter with a digital caliper. Test new materials in small prints before trying larger ones, and document everything while learning the

Figure GS-27   An impurity inside the filament, which can jam the extruder (*left*). A twisted length of filament (*right*). This will not pass through the extruder and will jam and possibly damage it.

filament's performance limits. You might also do a YouTube search for "3D printing filament reviews."

Responsible Waste Disposal

Some PLA is compostable, but putting PLA in a recycling bin isn't a good practice because it can biodegrade in the recycling process. Ask your local recycler if it takes printing scraps and, if so, what filament types. Sustainable filaments exist that are made from recycled materials such as potato chip bags, milk cartons, and beer by-products.

Supports are extra plastic that has to be disposed of, so avoid or minimize them. When they're unavoidable, use biodegradable, compostable, or dissolvable support filament.

There are filament recyclers that grind up PLA and ABS prints and leftover spool scraps. And there are filament extruders that turn those scraps, or pellets, into filament (Figure GS-28). These enable you to create your own colors (Figure GS-29).

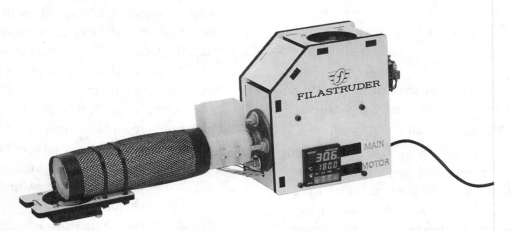

Figure GS-28 The Filastruder makes filament from pellets or recycled filament (filastruder.com).

Figure GS-29 Artist Nick Clark designed and printed these with filament made with the Filastruder. See more of his art at blog.nickclark.co.

All that said, keep a box of scraps (Figure GS-30) because they can be useful. For example, you might use an old raft to prop up prints whose supports failed (Figure GS-31). Scraps are also good for practicing post-processing techniques on.

SketchUp

SketchUp is used for some of this book's projects. Here is some information needed in all of them.

How to Download Extensions

Our SketchUp projects will require tools called *extensions*. Download and install them from SketchUp's Extension Warehouse. You'll need to make a free account with Google (accounts.google.com) to do this. Once your

Figure GS-30 Keep a box of scraps to practice post-processing techniques on.

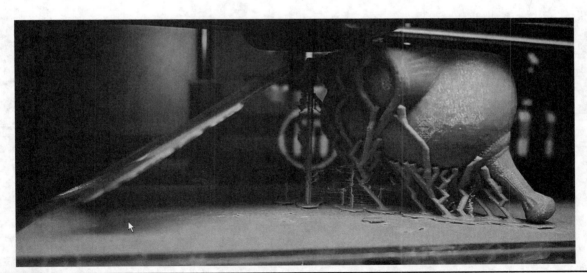

Figure GS-31 A raft from a previous print propped up this print when its supports failed.

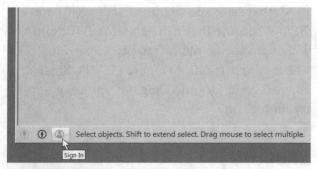

Select objects. Shift to extend select. Drag mouse to select multiple.

Sign In

Figure GS-32 Sign into your Google account here before going to the Extension Warehouse.

account is created, sign in through SketchUp by clicking on the icon in the lower-left corner of the workspace (Figure GS-32). Then click on the Extension Warehouse icon (Figure GS-33),

Extension Warehouse
Add extensions to SketchUp.

Figure GS-33 The Extension Warehouse icon.

and you'll be taken there (Figure GS-34). Search for an extension by typing its name into the Warehouse search field. Install each extension from its page by clicking the red button. That button will say *Download* when you're not signed in, *Install* when you are, and *Uninstall* when you're signed in and have already installed it. Figure GS-35 shows the page for the SketchUp STL extension.

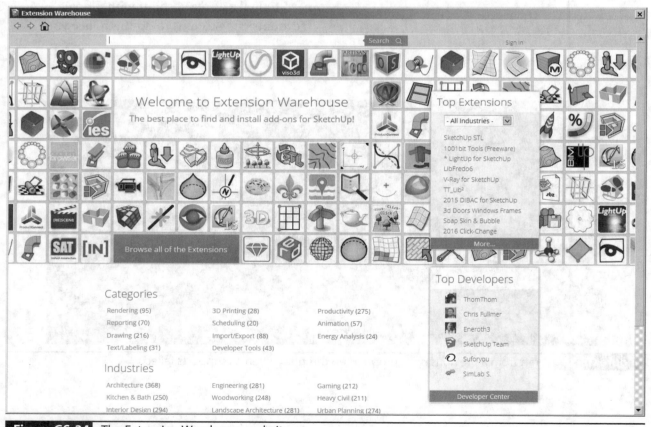

Figure GS-34 The Extension Warehouse website.

Figure GS-35   The SketchUp STL extension page.

Extensions Needed

The SketchUp STL, CleanUp³, and Solid Inspector² extensions are needed for every SketchUp project in this book. Other extensions will occasionally be needed too, and are listed in the "Things You'll Need" chart at the beginning of each project.

SketchUp STL imports STL files and exports SketchUp models as STL files. CleanUp³ erases stray edges, purges unused items, merges coplanar faces, and does whatever else it can to reduce the model to essential lines and faces. Solid Inspector² inspects grouped models and fixes problems that keep the group from being solid (watertight), a 3D printing requirement. Sometimes it fixes everything; other times it fixes what it can and flags what it can't.

When CleanUp³ or Solid Inspector² aren't enough to make a model solid, its problems may need to be manually fixed or fixed with other programs or websites. However, even if SketchUp doesn't consider a model solid, that model might still be printable. But the closer you can get it to solid, the greater the likelihood of being able to print it.

Resizing SketchUp Models

SketchUp models often lose their dimensions when exported as STLs, especially if they weren't modeled in millimeters. However, they do maintain their proportions, so their dimensions can be easily fixed in a slicer or other software program. Figure GS-36 shows the file from our armchair coaster project imported into Autodesk Meshmixer. Click on the Analysis icon, then on Units/Dimensions, and type the desired size into one of the dimension text fields. The other dimensions will adjust proportionately.

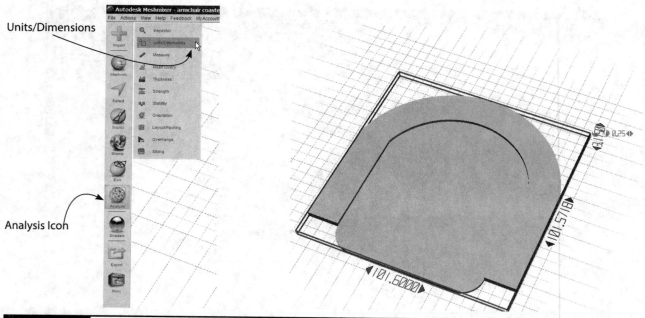

Units/Dimensions

Analysis Icon

Figure GS-36 Change the dimensions of a model inside Meshmixer with the Units/Dimensions function.

Websites and Programs That Check for Printability

It is prudent to run the model through a program that checks for flaws, especially if it will take many hours to print. Following are some programs and websites that check files and fix what they can. Know that automatic fixers just make the model printable; the model won't necessarily be fixed in a way that maintains its appearance. Fixing a model may involve running it through one of these apps and then opening the fixed model in the original program to work on it some more. Sometimes you can fix it in an automatic program, download the fixed STL, then run that STL through the program again for an even better result. You can also bounce an STL between multiple programs, fixing it in each for a better final result. Each program may return a different result on the same model, so try them all.

- *Netfabb.com.* This Autodesk program at netfabb.com/ inspects and repairs models, generates supports, and more. There are three subscription options: Standard, Premium, and Ultimate. You can download a 30-day free trial.

- *Netfabb Online (Azure) Service.* This free Autodesk website at https://service.netfabb .com/login.php checks and repairs STL files. You can log in with a Google account.

- *Tinkercad.* This free Autodesk app at tinkercad.com automatically repairs STL files on import. Its pay (monthly subscription) option allows commercial use.

- *Trimble SketchUp 3D Warehouse.* The model repository at 3dwarehouse.sketchup.com has a free Printables feature. Click the "I want this model to be Printable" checkbox when uploading a SketchUp model. The Warehouse sends it to its partner Materialise (materialise.com), where it is converted to an

STL file, analyzed, flaws fixed, and returned to the Warehouse for you to download. You can make any models you've already uploaded to the Warehouse printable by signing into your Warehouse account, going to the model's details page, clicking the Edit button, and scrolling to the Printables box at the bottom of the page.

- *Microsoft 3D Repair Service* (tools3d .azurewebsites.net). This free site fixes your file and returns it in an .3mf format. To open it, download a Microsoft program called 3D Builder, a program that also comes with the Windows 10 operating system.

- *Simplify3D* (simplify3d.com). This slicing program has some repair options.

- *Make Printable* (makeprintable.com). This free Web app repairs and checks files for 3D printability. You specify what kind of printer and material you're using and file type wanted.

- *Autodesk Meshmixer*. Download this free Autodesk program at meshmixer.com. Its Analysis feature inspects an STL file for defects and offers automatic and manual repair options. Meshmixer also lets you combine multiple models, hollow them out, and export in multiple formats.

Summary

In this chapter we discussed subjects relevant to all our projects: printer features, filament types, build plates and treatments for them, slicer settings, selling work, ways to be green, finding and adding extensions to SketchUp, and websites that check your file for printability. Let's start making things now!

Resources

- Website with lots of 3D printing information: http://3dpc.co/.

- 3D printer comparisons: http://makezine.com/comparison/3dprinters/.

- www.3ders.org/pricecompare/3dprinters/, and makezine.com/comparison/3dprinters/.

- Delta printers: http://builda3dprinter.eu/information/why-a-delta/.

- Extruder types: http://reprap.org/wiki/Hot_End_Comparison.

- Bowden versus direct extruders: www.fabbaloo.com/blog/2015/11/11/bowden-or-direct-a-primer-on-extruder-styles.

- Charts of popular and exotic filament types: https://pinshape.com/blog/popular-3d-printing-filaments-3d-printer-filament-types/ and https://all3dp.com/best-3d-printer-filament-types-pla-abs-pet-exotic-wood-metal/.

- Popular filament sources: Inland brand from MicroCenter (microcenter.com), Hatchbox (hatchbox3d.com), MatterHackers (matterhackers.com), PrintedSolid (printedsolid.com), Proto-pasta (proto-pasta.com), MakerBot (makerbot.com) Colorfabb (colorfabb.com), Esun (esun3d.net), 3D Supply Guys (www.3dsupplyguys.com/), Protoparadigm (protoparadigm.com), and MakerGeeks (makergeeks.com).

- PrintInZ products: www.printinz.com/.

- Sources for food-safe filament and sealants: bnktech.kr, formfutura.com, germanreprap.com, and masterbond.com/.

- Sources for sustainable filaments: bioinspiration.eu, 3dbrooklyn.com/, 3dfuel.com/, and 3domusa.com.

- Download 3D Builder software: microsoft .com/en-us/store/apps/3d-builder/ 9wzdncrfj3t6.
- Print shredder: filamaker.eu/.
- Comparison of filament extruders: www.aniwaa.com/blog/the-best-filament -extruders-for-3d-printers/.
- Service bureaus that print and sell your designs: Shapeways, Ponoko, and Materialise.
- Model repositories: 3DWarehouse.sketchup .com, youmagine.com, grabcad.com, and pinshape.com.
- Download Octoprint: octopi.octoprint.org/.

Architectural Symbol Coaster

PROBLEM: An interior designer wants some coasters that look like architectural symbols to fit with her whimsical office decor. We'll model the plan (top) view of an armchair symbol using AutoCAD and SketchUp Pro, slice it in MakerBot Desktop, and print it in two colors on a single-extruder MakerBot Replicator 2.

Find an Existing 2D Drawing

Let's find an existing 2D DWG (AutoCAD) drawing to model, as this is often quicker than modeling an item from scratch. A search for "chair" in the free CAD library at Cadforum. cz includes the DWG file shown in Figure 1-1. Download it and open in AutoCAD (Figure 1-2).

Things You'll Need

Description	Source	Cost
Computer and Internet access	Your own, or one at a makerspace	Variable
Google account	accounts.google.com	Free
Autodesk account	Autodesk.com	Free
CADforum account	Cadforum.cz	Free
SketchUp Pro software	Sketchup.com	Free trial or $690
Four SketchUp extensions: SketchUp STL, CleanUp^3, Solid Inspector^2, and S4U Make Face	SketchUp Extension Warehouse	Free
Autodesk AutoCAD software	Autodesk.com	Free trial or variable subscription fee
3D printer and slicing software	Your own, one at a public makerspace, or one at an online service bureau	Variable
Thumb drive (needed only for offsite printing)	Computer or electronics store	< $10
Spool of PLA filament	Amazon, Microcenter, online vendor	Variable

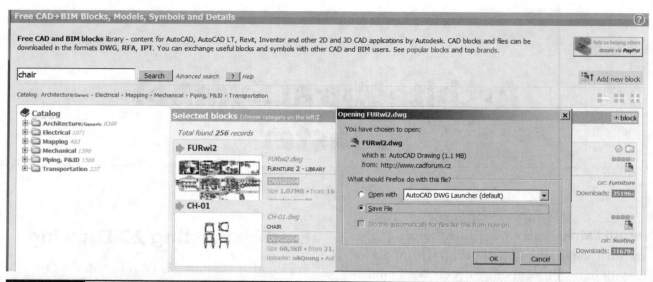

Figure 1-1 Download and save the DWG chair file from Cadforum.cz.

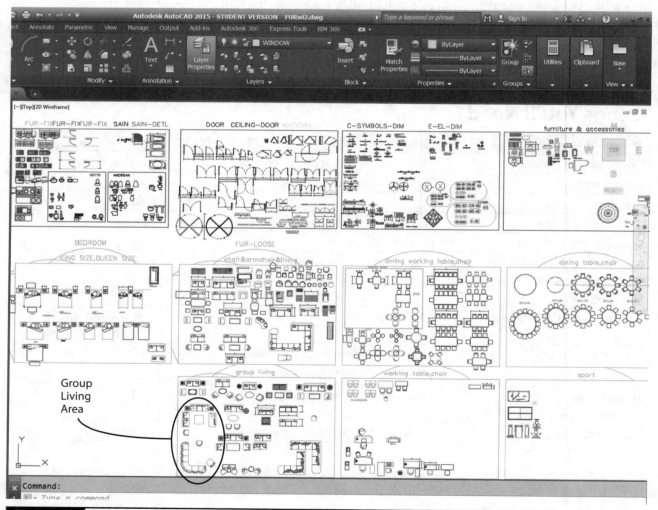

Figure 1-2 The entire file opened in AutoCAD.

Edit the Drawing

Locate the chair shown in Figure 1-3 in the "group living" area, select it, right-click, and choose Clipboard/Copy. Then Click the New icon/Drawing/acad-Named Plot Styles template (Figure 1-4), right-click on the workspace, choose Clipboard/Paste to insert the chair, and click an insertion point (Figure 1-5). The chair will appear.

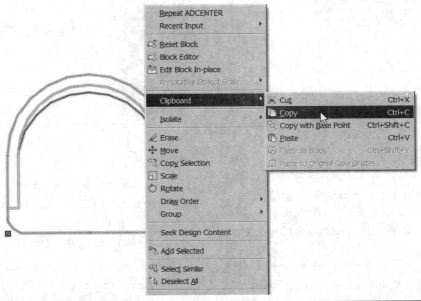

Figure 1-3   Select and copy this chair onto the clipboard.

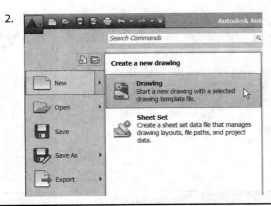

Figure 1-4   Open a new AutoCAD file.

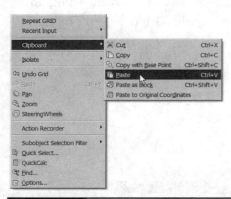

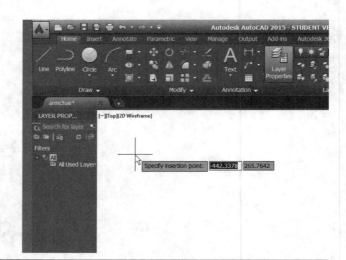

> The AutoCAD workspace might show a grid by default. You can turn it off by typing "GRID" in the command line and then double-clicking the Off option.

The chair is a block, meaning that all its lines are grouped together to act as one. We can't edit a block, so type *EXPLODE*, and then click on the block to select it. This will ungroup its lines.

This particular file has nested blocks, so you'll need to select and explode them individually (Figure 1-6). Eventually, you'll be left with a drawing that looks like Figure 1-7.

This drawing has gaps between some lines, so fill them in by typing *LINE* and connecting the endpoints. Now the chair is good enough to use. If you know how to use AutoCAD, you can edit the file to make it more to your liking, such

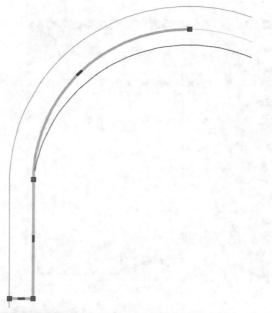

Figure 1-6   The highlighted curve and lines are a nested group.

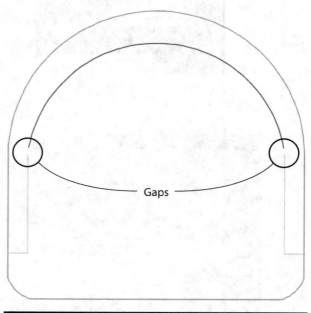

Figure 1-7   The completely exploded chair.

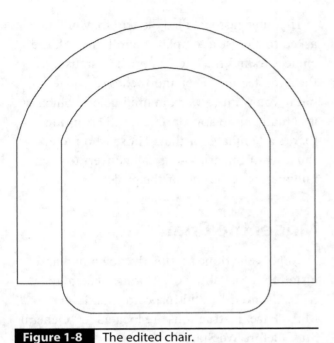

Figure 1-8 The edited chair.

as adjusting the curve or making the backrest thicker. Figure 1-8 shows the edits I made to the chair.

Import the Chair File into SketchUp Pro

The ability to import a DWG file into SketchUp is a Pro feature. Launch SketchUp Pro, click File/Import, select AutoCAD Files in the Files of Type field, and navigate to the file (Figure 1-9). Before importing it, click on the Options button. Then click the *Merge coplanar faces* and *Orient faces consistently* boxes (Figure 1-10). If the file's correct units aren't shown in the Units field, click the dropdown arrow and choose the correct one. Click OK and hit the ENTER key to import the DWG file into SketchUp.

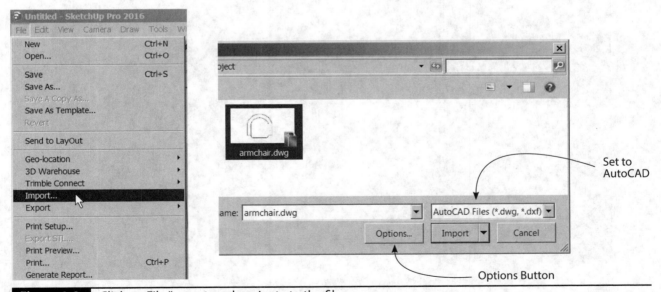

Figure 1-9 Click on File/import, and navigate to the file.

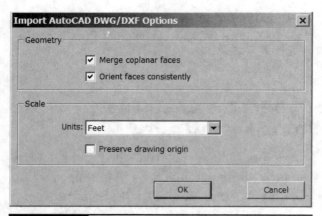

Check the upper two boxes.

A small message box will appear telling you what has been imported. Click that box closed and the chair will appear in the workspace (Figure 1-11).

If you've just installed SketchUp, you'll be asked to choose a template. Scroll through the choices to find the one you prefer. I'm using the Architectural Feet and Inches template, but if you want to work in millimeters, which is actually more appropriate for 3D printing, there's a template for that. Those who prefer thinking in Imperial units can convert to millimeters at the end of the project.

Model the Chair

If you haven't done so already, download and install four SketchUp extensions: SketchUp STL, CleanUp[3], Solid Inspector[2], and S4U Make Face. Instructions are located in "General Stuff Before We Start."

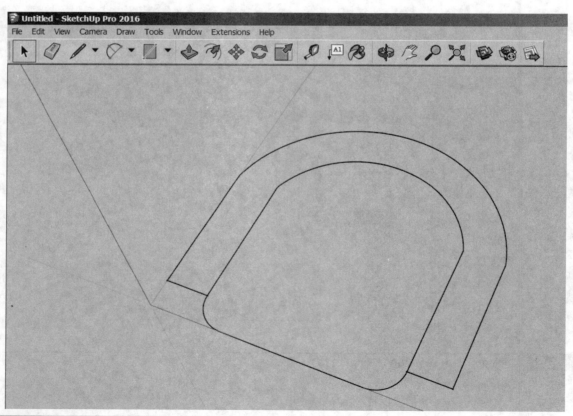

Figure 1-11   The imported DWG file.

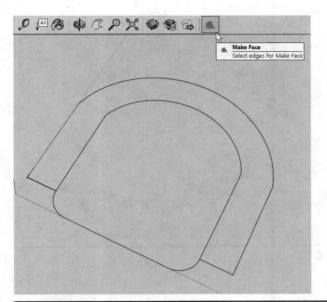

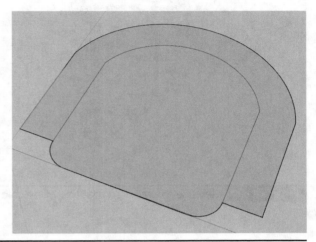

Figure 1-12   Select the chair and click the Make Face tool to fill it in.

The drawing is just an outline; it needs faces. There are a couple of ways to make them. One is to trace a line on the armrest and a line on the cushion. That should fill both parts in. Another is to select the whole chair by dragging the cursor around it and clicking the Make Face extension tool. That will fill in the whole chair (Figure 1-12). If the S4U Make Face tool doesn't appear when you install it, right-click on the toolbar, and a list of all available tools will appear (Figure 1-13). Find the tool you want and click its checkbox to make its icon appear.

Note that the chair's faces are gray. That's actually the back; SketchUp often gets the faces turned around. Simply click on the faces to highlight them (hold the SHIFT key down to make multiple selections), right-click, and choose Reverse Faces. This will flip them over so that the front, or white face is up (Figure 1-14). Having faces positioned properly is important to make the file 3D printable.

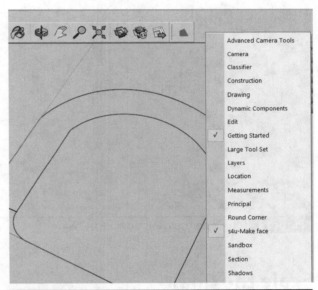

Figure 1-13   Right-click on the toolbar to see a list of all tools. Check the ones you want to use.

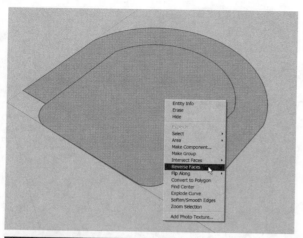

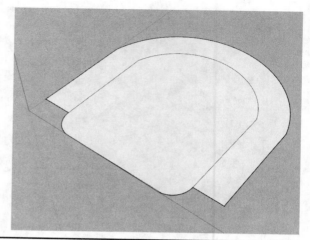

Figure 1-14 Reverse the faces so that the white color is up.

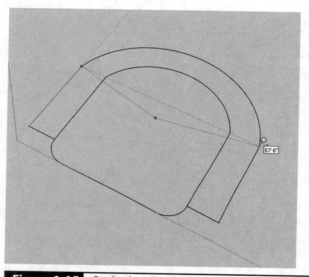

Figure 1-15 Scale the chair down to 4" wide.

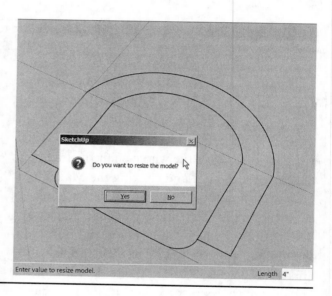

Scale the Chair

Let's make the chair 4" wide. Click the Tape Measure tool on opposite ends to see its current size. Figure 1-15 shows that its width is over 67 feet. Immediately type *4"*. You'll be asked if you want to rescale; click OK. The chair will scale to that size. Because it will now appear tiny, click the Zoom Extents tool to see it (Figure 1-16).

Figure 1-16 Click Zoom Extents to fill the screen with the chair.

Add Volume

Now turn this 2D drawing into a 3D model. Click the Push/Pull tool (Figure 1-17), hover it over the seat, move it up 1/16", and click to place. Push/pull the armrest up 1/8" (Figure 1-18). Hold the CONTROL key down (COMMAND key on a Mac) while push/pulling so that the underside of the chair doesn't get lifted with the top and disappear. If you do forget to hold the CONTROL/COMMAND key down and the underside of the chair disappears, just trace a line with the Pencil tool to restore it (Figure 1-19).

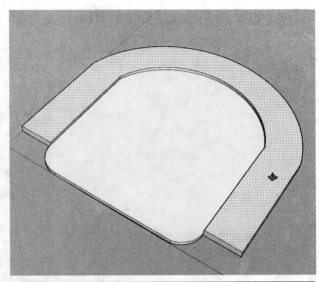

Figure 1-18 Push/pull the chair faces up.

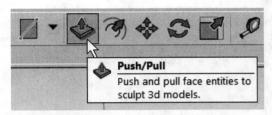

Figure 1-17 Push/Pull tool.

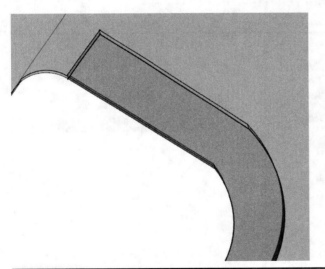

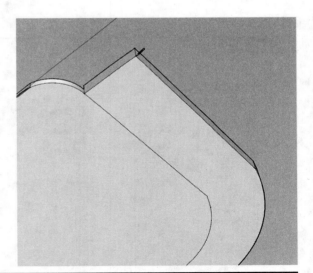

Figure 1-19 You can restore a face by tracing one of its edges with the Pencil tool.

Check the Model for Problems

The CleanUp[3] and Solid Inspector[2] extensions check the model for problems. Click Extensions/ CleanUp[3]/Clean (Figure 1-20). Next, click Tools/ Solid Inspector[2]. Here Solid Inspector[2] found and was able to fix all problems (Figure 1-21).

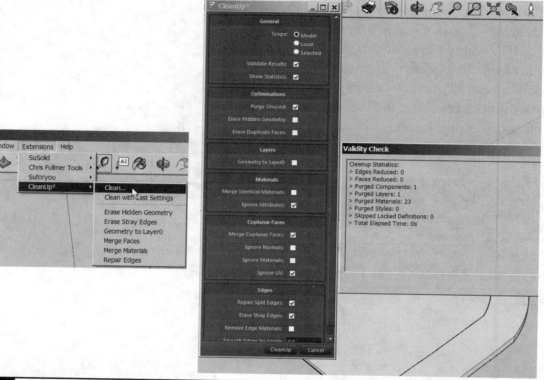

Figure 1-20   Run CleanUp[3] on the model.

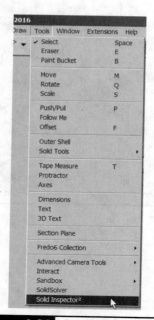

Figure 1-21   Run Solid Inspector[2] on the model.

Export It as an STL File

Now it's time to use the SketchUp STL extension (Figure 1-22). Click File/Export STL, and in the window that appears, choose the model's units and Binary (it's a smaller file than ASCII). Click Export, and we're almost ready to print.

Print It!

Import the STL file into your slicer. Figure 1-23 shows the MakerBot Desktop settings. Figure 1-24 shows the coaster under construction and the final print. It has a 10 percent infill and was printed with one extruder on an acrylic bed covered in painter's tape wiped with isopropyl alcohol.

To create a two-color coaster with MakerBot Desktop and one extruder, watch the coaster while it's under construction and pause it after the cushion is completed (Figure 1-24). Then choose Change Filament/Unload (Figure 1-25). Physically unload the filament, and then choose Change Filament/Load. Physically load the new filament, and then choose Resume Build. The armrests will print in the new color.

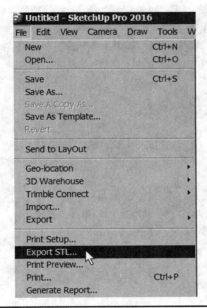

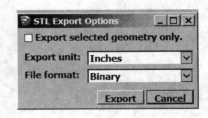

Figure 1-22 Export the model as an STL file.

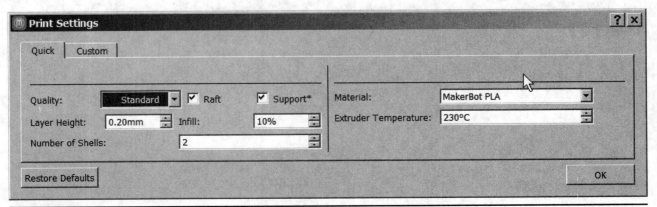

Figure 1-23 MakerBot Desktop settings.

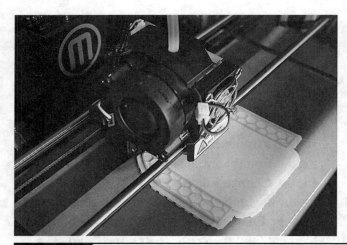

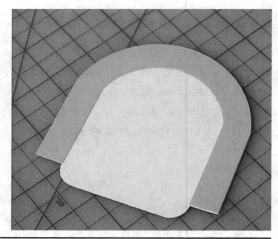

Figure 1-24   The coaster under construction and the finished print.

Figure 1-25   Pause the MakerBot either directly or through the MakerBot Desktop slicer.

Phone Case

PROBLEM: A game fan wants to interpret the game's motif onto a phone case. In this project we'll download a premade case, edit it in SketchUp, make it 3D printable in Meshmixer, slice it with MakerBot Desktop, and print it in PLA with a MakerBot Replicator 2.

Find a Phone Case File and Import It into SketchUp

Use the yeggi.com search engine to find an already-made case for your phone. Figure 2-1 shows a file that I found for mine. Sometimes the designers include source files, that is, the model

Things You'll Need

Description	Source	Cost
Computer and Internet access	Your own or one at a makerspace	Variable
Google account	accounts.google.com	Free
Autodesk account	Autodesk.com	Free
SketchUp make software	Sketchup.com	Free trial or $690
Three SketchUp extensions: SketchUp STL, CleanUp[3], and Solid Inspector[2]	SketchUp Extension Warehouse	Free
Autodesk Meshmixer software	Meshmixer.com	Free
3D printer and slicing software	Your own, one at a public makerspace, or one at an online service bureau	Variable
Thumb drive (needed only for offsite printing)	Computer or electronics store	< $10
Spool of PLA filament	Amazon, Microcenter, or online vendor	Variable

Figure 2-1   A phone case file found on the Web.

in the original program. Those are nice because they make editing easier, assuming that you have the program it was made in and know how to use it.

To import an STL file, click File/Import (Figure 2-2). Click the dropdown arrow in the file's text field, and set it to STL. Click the Options button, and check Merge coplanar faces. Set the Units to the STL file's units. If you don't know what they are, that's okay. Choose millimeters (mm) as a default; you can resize the file inside SketchUp if necessary. Then navigate to the file, and hit ENTER. The file may appear very small, so click Zoom Extents (Figure 2-3) to fill the screen with it (Figure 2-4).

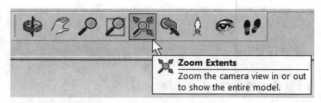

Zoom Extents
Zoom the camera view in or out to show the entire model.

Figure 2-3   The Zoom Extents tool.

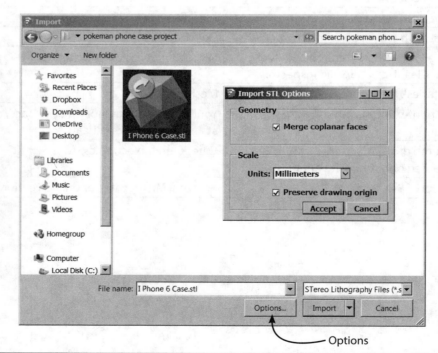

Figure 2-2   Importing the STL file into SketchUp.

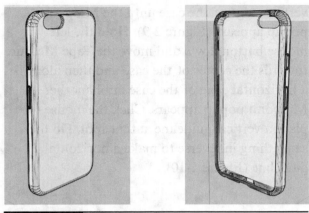

Figure 2-4   The imported phone case.

The phone case looks this clean because we checked the Merge coplanar faces button. If we hadn't, it would have imported with a lot

of polygon lines. If you forget to check that button, run the CleanUp[3] extension (Extensions/CleanUp[3]/Clean) to get rid of them.

Open More Toolbars

SketchUp's workspace opens with the Getting Started toolbar. Click on View/Toolbars to see more toolbars (Figure 2-5). Check the boxes in front of Large Toolset, Standard, and Views. Then close the screen. Note the Undo arrow on the Standard toolbar (Figure 2-6) because it is a handy tool.

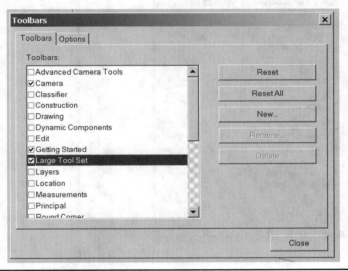

Figure 2-5   View/Toolbars shows all available tools.

Figure 2-6   The Undo icon.

Sketch a Design

Click Camera/Parallel Projection. This displays the model orthographically, which makes sketching easier. Then click the View toolbar's Back icon (Figure 2-7). Now you're ready to draw the motif.

We need two guidelines that intersect at the case's center. Click on the Tape Measure tool (Figure 2-8), and then run it slowly along a vertical edge of the case until the Midpoint popup appears (Figure 2-9). Hold the left mouse button down and move the Tape Measure towards the center of the case and then along a horizontal edge of the case until another Midpoint popup appears. Click the mouse to place a vertical guideline at that spot. Do the same thing in reverse to make a horizontal guideline (Figure 2-10).

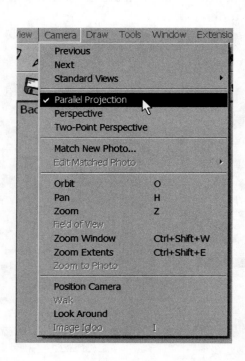

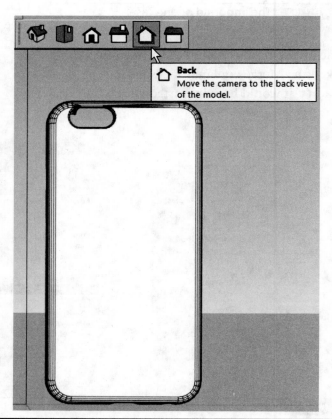

Figure 2-7   The phone case's backside displayed orthographically.

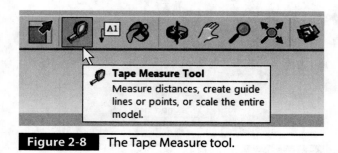

Figure 2-8   The Tape Measure tool.

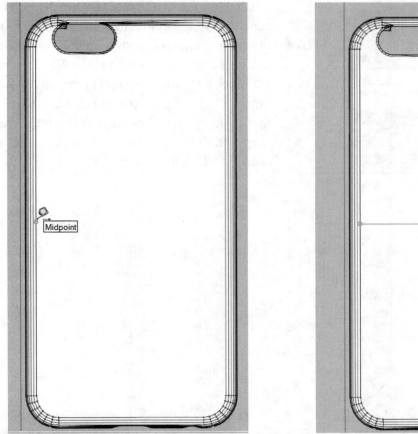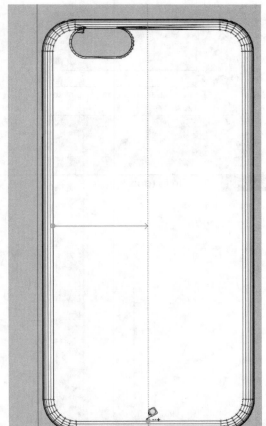

Figure 2-9 Draw a vertical guideline.

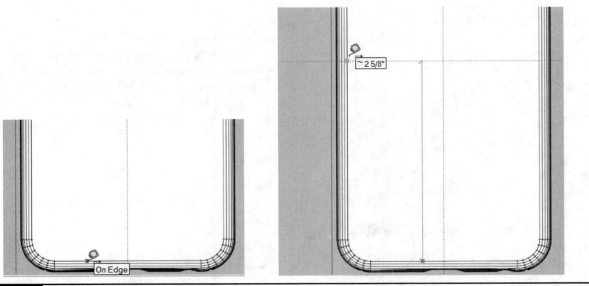

Figure 2-10 Draw a horizontal guideline.

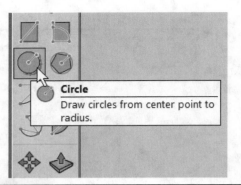

Figure 2-11 The Circle tool.

Click on the Circle tool (Figure 2-11). Then click it onto the guidelines' intersection, move the cursor until 1/4" shows up in a popup (or just type *1/4*), and click to place. Repeat with 1/2" and 3/4" radius circles (Figure 2-12). Look for the On Face popup while drawing to ensure that the circle sketches are indeed on the case's face. You don't need the guidelines anymore, so click Edit/Delete Guides.

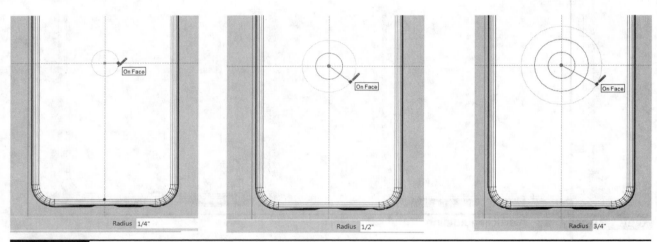

Figure 2-12 Draw three circles with 1/4", 1/2", and 3/4" radii.

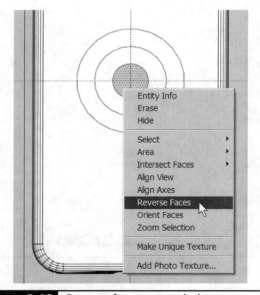

Figure 2-13 Reverse faces as needed.

Tip: If faces become reversed—that is, the gray back is up instead of the white front—just select it, right-click, and choose Reverse Faces (Figure 2-13).

Click on the Rectangle tool (Figure 2-14), and draw a rectangle from edge to edge, as shown in Figure 2-15. Then click the Eraser tool (Figure 2-16) on overlapping lines to make the sketch look like Figure 2-17.

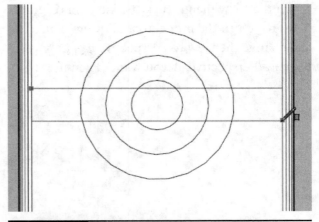

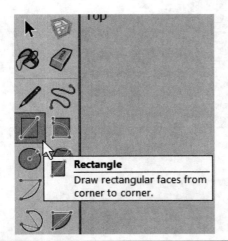

Figure 2-14   The Rectangle tool.

Figure 2-15   Draw a rectangle over the circles.

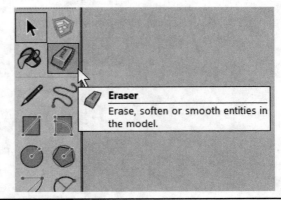

Figure 2-16   The Eraser tool.

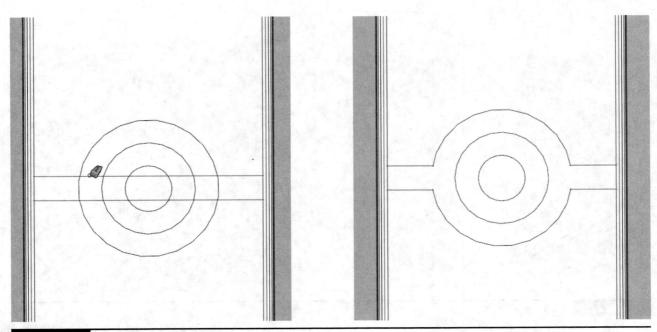

Figure 2-17   Erase overlapping lines.

On the View menu, click the Iso icon (Figure 2-18) to return the model to a 3D view. You can change the view to perspective by clicking Camera/Perspective, if you want. Then use the Push/Pull tool (Figure 2-19) to pull the outer ring up 1/16", the button up 1/8", and the top and bottom halves up 1/8" (Figure 2-20).

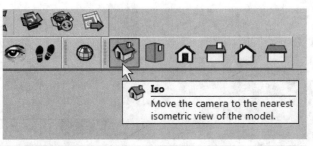

Figure 2-18   The Iso icon.

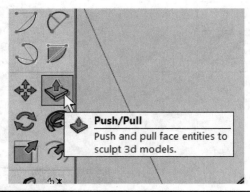

Figure 2-19   The Push/Pull tool.

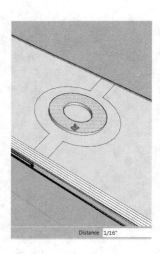
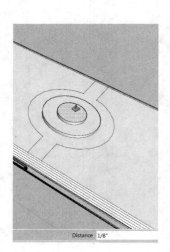
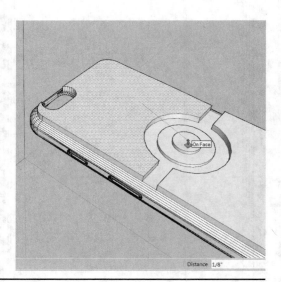

Figure 2-20   Pull the sketches up.

Optional: Paint and Export as an OBJ File

If you plan to send this model to an online service bureau such as Shapeways.com, you can add colors and textures because their commercial machines can print them. Click on the Paint Bucket tool (Figure 2-21),

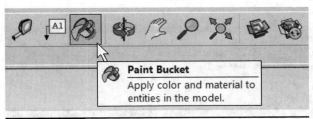

Figure 2-21 The Paint Bucket tool.

scroll through the Colors folder, click on the appropriate colors, and click them onto the model. Or click on the Patterns folder, and click a fun pattern onto it (Figure 2-22). You can also import your own pattern.

Export the model as an OBJ file (a SketchUp Pro feature) because that format holds color information. Click on File/Export/3D Model, and then scroll to OBJ file (Figure 2-23). When you export it, two files will be made: the OBJ file and an MTL file (sometimes a JPG file will appear, too). The MTL file contains the color and pattern (Figure 2-24). Keep all those files together in their own folder when uploading to a service bureau or when importing into another program.

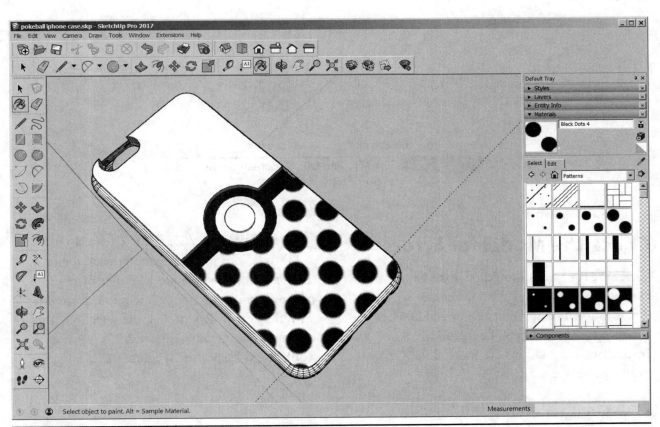

Figure 2-22 Paint colors or textures onto the model.

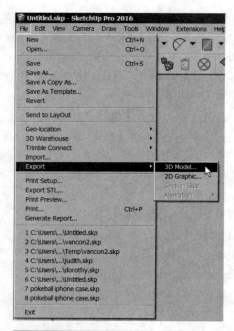

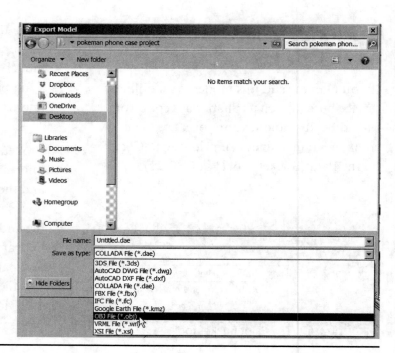

Figure 2-23 Export the model as an OBJ file.

Untitled.mtl   Untitled.obj

Figure 2-24 The MTL and OBJ files created when exporting the model.

Check the Model for Errors

Checking for errors should be done before exporting the model. If you haven't done so already, download and install three SketchUp extensions: SketchUp STL, CleanUp[3], and Solid Inspector[2]. Instructions are located in "General Stuff Before We Start."

Click Extensions/CleanUp[3]/Clean (Figure 2-25) to let this tool find and fix what it can. Then select the whole model either by triple-clicking on it or dragging a window around it with the cursor. Right-click the selection, and choose Make Group (Figure 2-26). Then click Tools/Solid Inspector[2] (Figure 2-27).

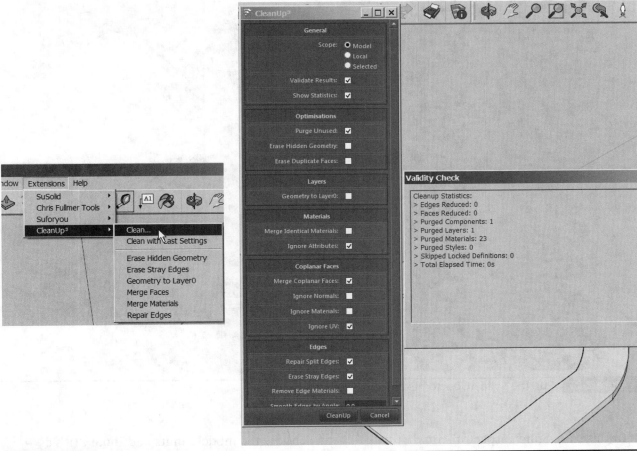

Figure 2-25   Run the CleanUp³ extension.

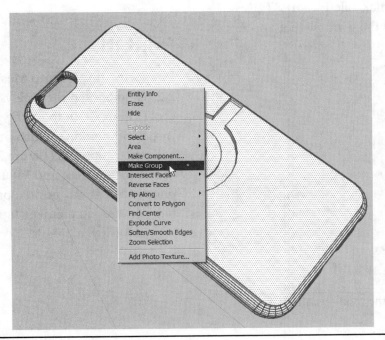

Figure 2-26   Turn the model into a group.

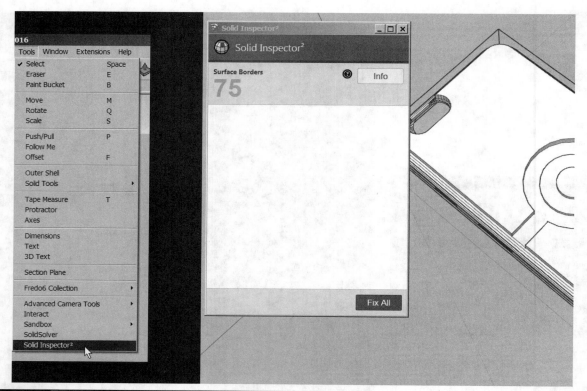

Figure 2-27 Run the Solid Inspector² extension.

In this case, Solid Inspector² couldn't fix everything, so the model isn't watertight, but we'll continue, anyway. When finished, export the file as an STL. Click File/Export STL (Figure 2-28), and in the window that appears, choose the model's units and Binary (it's a smaller file than ASCII). Click Export, and you're almost ready to print.

Print It!

I imported this file into Meshmixer (Figure 2-29) and clicked the Analysis icon/Inspector tool (Figure 2-30) to fix the problems that Solid Inspector² couldn't. I also oriented the case and generated supports (Figure 2-31). Figure 2-32 shows the settings, and Figure 2-33 shows the phone case. It was printed on a cold acrylic build plate covered in painter's tape and wiped down with isopropyl alcohol.

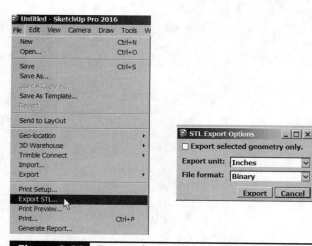

Figure 2-28 Export the model as an STL file.

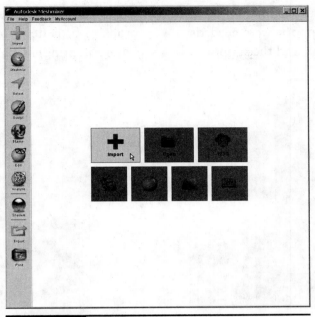

Figure 2-29 Import the phone case into Meshmixer.

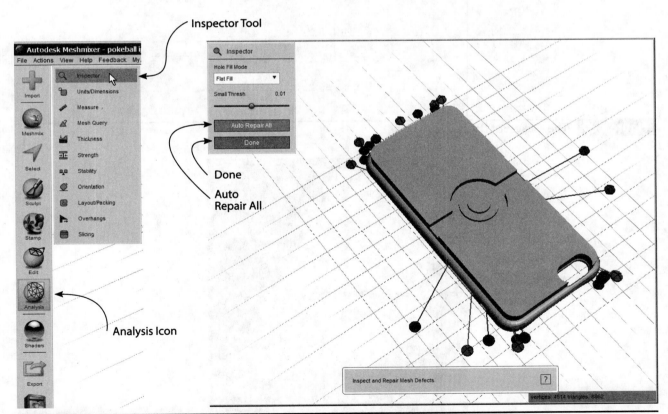

Figure 2-30 Analysis/Inspector finds problems. Fix them by clicking *Auto Repair All* and then *Done*.

The print orientation isn't optimal for strength, but too many supports were needed when the print was laid flat, and the supports failed when the print was angled. This would probably be better printed on an SLS (Selective Laser Sintering) machine.

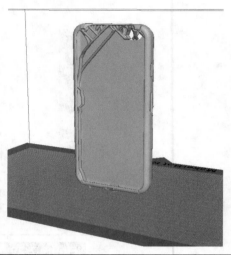

Figure 2-31   The phone case with supports and positioning generated by Meshmixer.

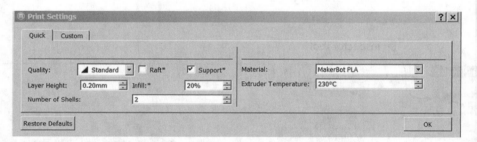

Figure 2-32   The MakerBot Desktop settings.

Figure 2-33   The printed phone case.

Guardian Lion Bank

PROBLEM: A mom hopes that a fun bank shaped like a guardian lion will encourage her kids to save their allowance. In this project we'll download a ready-made model from Thingiverse, edit it in Meshmixer, slice it with MakerBot Desktop, and print it in PLA with a MakerBot Mini.

Find a Model

At Thingiverse.com, search for "Guardian Lion." Then download the model shown in Figure 3-1, which happens to be a reality capture scan. Here's the link: www.thingiverse.com/thing:659634. Import it into Meshmixer (Figure 3-2).

Things You'll Need

Description	Source	Cost
Computer and Internet access	Your own or one at a makerspace	Variable
Autodesk account	Autodesk.com	Free
Autodesk Meshmixer software	Meshmixer.com	Free
3D printer and slicing software	Your own, one at a public makerspace, or one at an online service bureau	Variable
Thumb drive (needed only for offsite printing)	Computer or electronics store	< $10
Spool of PLA filament	Amazon, Microcenter, or online vendor	Variable

Figure 3-1 Guardian Lion model.

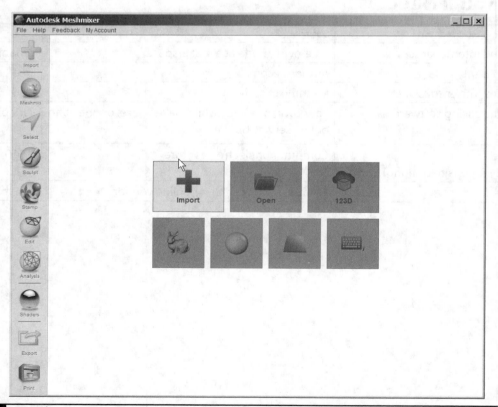

Figure 3-2 Click the Import button to bring the model into Meshmixer.

Hollow the Model

The model is solid, so hollow it out (Figure 3-3). Click on Edit/Hollow. The dialog box that appears shows an offset (wall thickness) default of 2 mm and the model becomes transparent so that you can see the offset. Click on the 2 mm entry to make a text box appear, and type *3 mm* to make the walls a bit thicker. Then click Accept.

Cut a Coin Slot

Click on the Meshmix icon and scroll through the Primitives menu to the box. Drag it into the workspace and let go (Figure 3-4). A Transform tool will appear over it. This has arrows, angles, and squares colored for each axis. Click and drag the white cube at the origin to make the whole box smaller. Click and drag the blue square to make the box narrower, and click the green square to make it shorter. The new dimensions appear both in the workspace and in the dialog box (Figure 3-5).

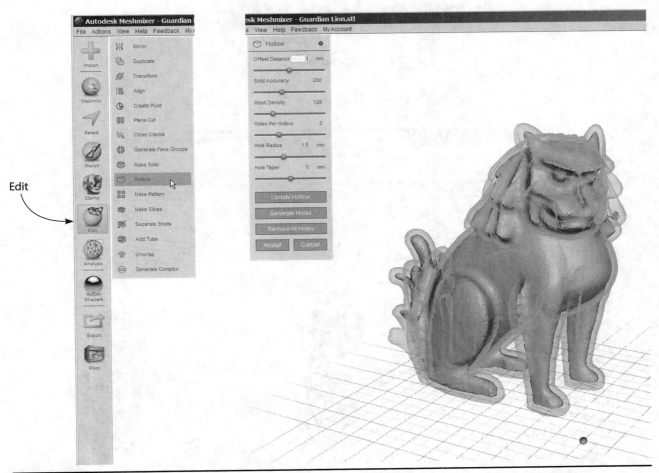

Figure 3-3   Hollow the model out.

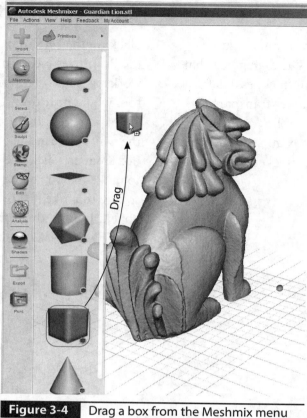

Figure 3-4 Drag a box from the Meshmix menu into the workspace.

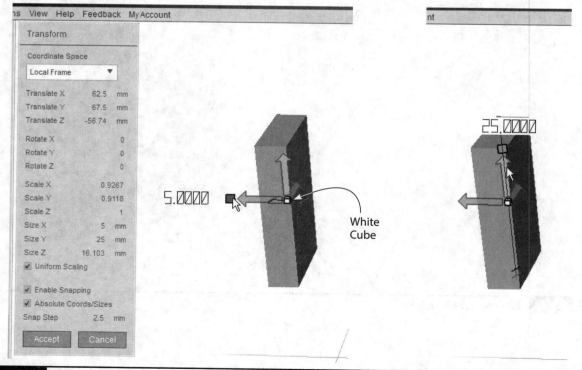

Figure 3-5 Alter the box's dimensions with the Transform tool.

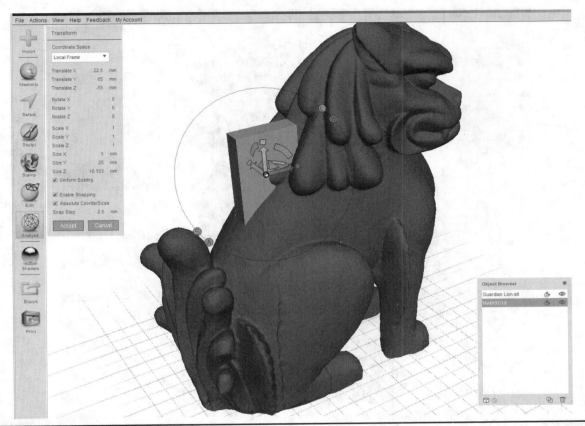

Figure 3-6 Position the altered box where you want the coin slot to be.

Position and insert the altered box where you want the coin slot to be (Figure 3-6), and then click Accept. You can display the model orthographically at View/Orthographic View (Figure 3-7) to make positioning easier. Then subtract the box from the lion by selecting Guardian Lion FIRST in the Objects Browser box, holding the SHIFT key down, and selecting MeshS018 (the name of the box) SECOND. Both are now highlighted, and a menu with Boolean options appears (Figure 3-8).

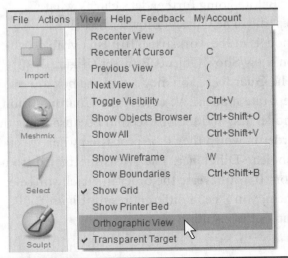

Figure 3-7 View the model orthographically to make positioning easier.

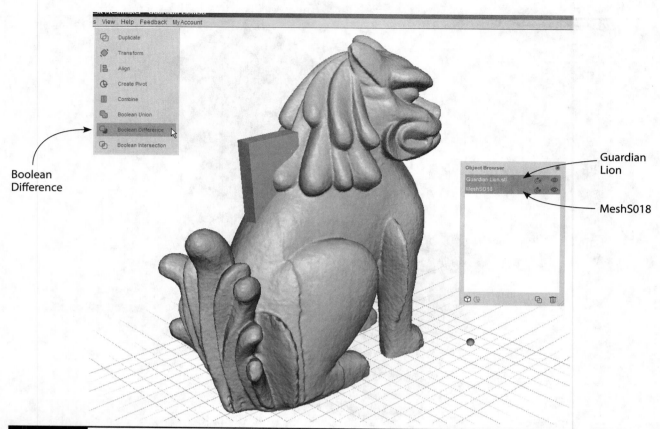

Boolean
Difference

Guardian
Lion

MeshS018

Figure 3-8 When both models are selected in the Objects Browser box, a Boolean menu appears.

Before going further, let's clarify what we just did. The Objects Browser box shows all models in the workspace. If it isn't visible, click on View/Show Objects Browser (Figure 3-9). The Guardian Lion model and the box are two separate models. We clicked the Guardian Lion model first and the box second because that's the order in which they'll be processed when we click Boolean Difference. The box is subtracted from lion. If you reverse the order in which you select them, you'll get a different result. Figure 3-10 shows the resulting coin slot.

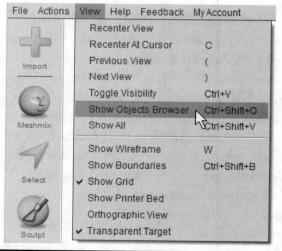

Figure 3-9 If the Objects Browser box doesn't appear, turn it on at View/Show Objects Browser.

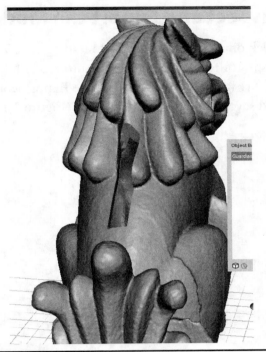

Figure 3-10 The coin slot.

Cut a Hole at the Bottom

Now we need an opening to retrieve stored coins. Open the Meshmix parts bin again and drag a cylinder out (Figure 3-11). Instead of dropping it on the workspace, drop it onto the model and position it (Figure 3-12). You can alter its size by moving the Dimension slider in the dialog box. Then extend the cylinder into the lion by moving the Offset slider. Click the dropdown arrow in the Composition Mode text field, and choose Boolean Subtract (Figure 3-12). The cylinder will subtract from the lion, leaving a hole (Figure 3-13).

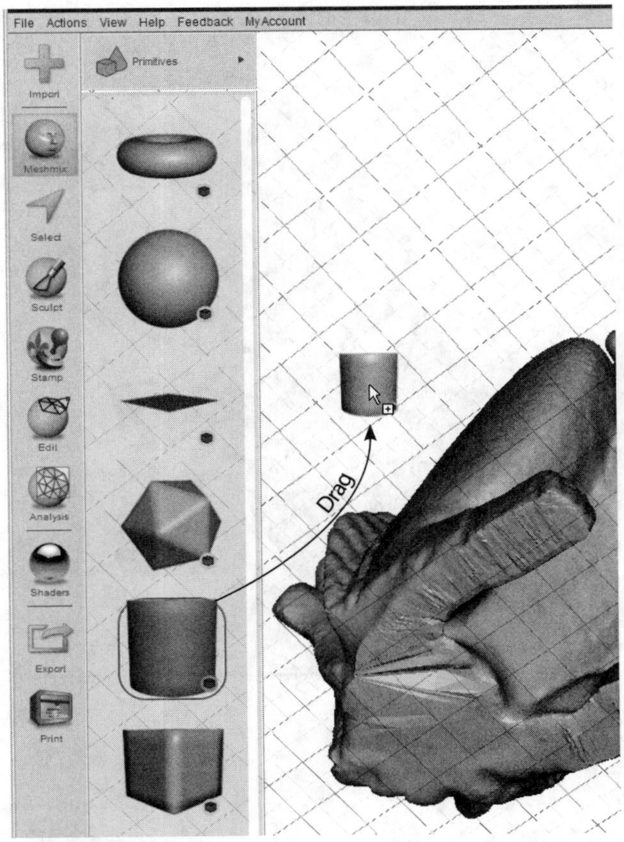

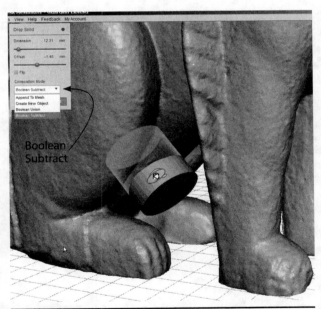

Figure 3-12 Drop the cylinder onto the bottom of the model, adjust and position it, and choose Boolean Subtract.

Figure 3-11 Drag a cylinder out of the Meshmix parts bin.

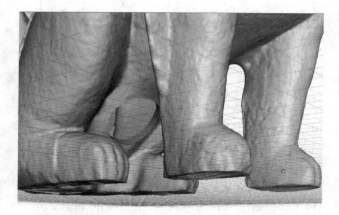

Figure 3-13   Subtract the cylinder from the lion model to make a hole at the bottom.

Inspect and Export the Model

Click on Analysis/Inspector to find defects. The Inspector found one; click on Auto repair all to fix it (Figure 3-14). Then click the Export icon and save the model as an STL file (Figure 3-15).

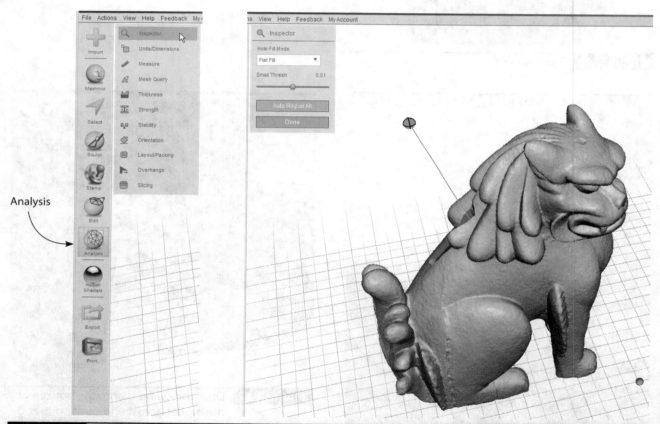

Figure 3-14   The pin shows a flaw.

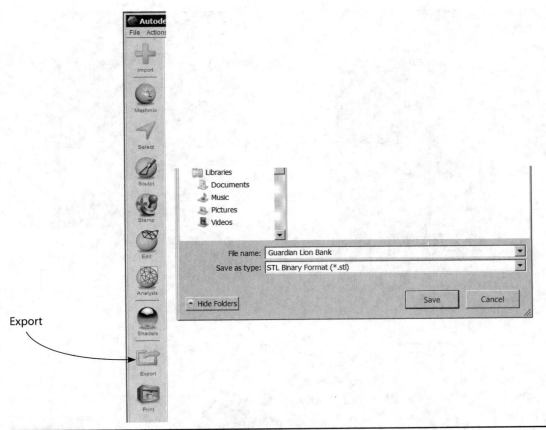

Export

Figure 3-15　Export the model as an STL file.

Try to Print It!

This model was a challenge to print, as models made from scans often are. First, I sliced it in Simplify3D and sent the model to the Taz 6. It failed after about half an hour, depositing stringy layers of filament on top of good layers and then "air printing." I sliced it in Cura and got the same stringy filament result. I tried generating supports in Meshmixer, sliced the file in Simplify3D, and sent it to the Taz again. This time the Taz spent so long trying to read the g-code that I quit the print operation, concluding that the interior structure of this model was too complicated.

Print It!

Rather than abandon the file, I returned it to Meshmixer and reduced its polygon count. This is done by clicking on the Select icon, clicking a spot on the model, clicking Modify/Select All, and then clicking Edit/Reduce (Figure 3-16). A dialog box will appear; type *95* in the Percentage field to reduce the polygon count by 95 percent (Figure 3-17). You can see a noticeable difference now, and the file size is much smaller (reduced from 35 to 1.5 MB), but the shape and features are still preserved. Export as an STL file.

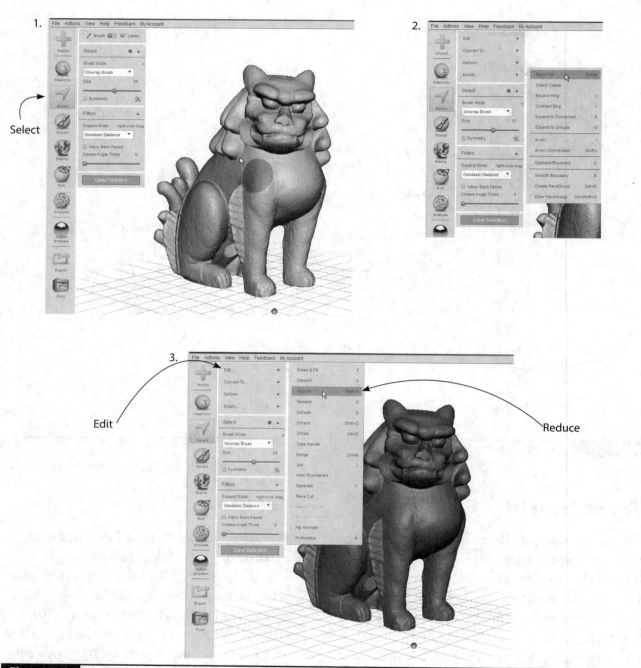

Figure 3-16 Select the entire model, and then click Edit/Reduce.

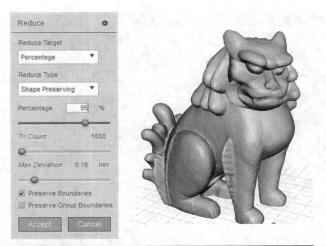

For added measure, I imported the STL file into www.tinkercad.com because Tinkercad fixes STL files on import. Then I exported it as an STL file (Figure 3-18), sliced it in MakerBot Desktop (Figure 3-19) and printed a miniature on a MakerBot Mini (Figures 3-20 and 3-21). A miniature is always prudent when you're unsure how a file will print before committing time and materials on a larger print. Success (Figure 3-20)! Search for a plug on Amazon to cover the coin slot, and enjoy your bank.

Figure 3-17 The file polygon count is reduced by 95 percent.

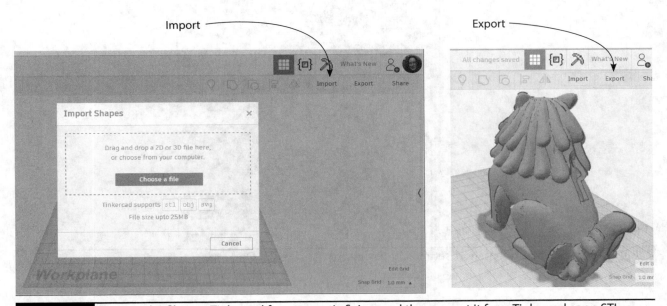

Figure 3-18 Import the file into Tinkercad for automatic fixing, and then export it from Tinkercad as an STL.

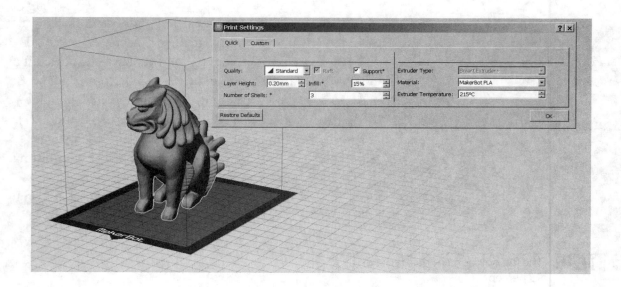

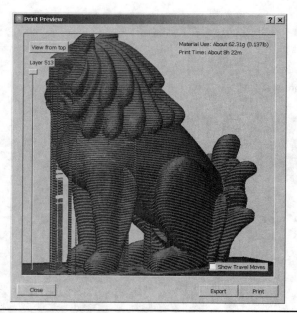

Figure 3-20 The printed bank with its supports.

Figure 3-21 Views after the supports were removed.

Art Stencil

PROBLEM: A middle school teacher plans to demonstrate stencil painting and thinks students will be more engaged if they can design their own. In this project we'll use SketchUp Make and Meshmixer to make a smiley face stencil, slice it in MakerBot Desktop, and print it in PLA on a MakerBot Replicator 2.

Open More Toolbars

SketchUp's workspace opens with the default Getting Started toolbar. Click on View/Toolbars to see more toolbars (Figure 4-1). Check the boxes in front of Large Toolset, Standard, and Views. Then close the screen. Note the Undo

Things You'll Need

Description	Source	Cost
Computer and Internet access	Your own or one at a makerspace	Variable
Google account	accounts.google.com	Free
SketchUp Pro software	Sketchup.com	Free trial or $690
Three SketchUp extensions: SketchUp STL, CleanUp³, Solid Inspector²	SketchUp Extension Warehouse	Free
Autodesk Meshmixer software	Meshmixer.com	Free
3D printer and slicing software	Your own, one at a public makerspace, or one at an online service bureau	Variable
Thumb drive (needed only for offsite printing)	Computer or electronics store	< $10
Spool of PLA filament	Amazon, Microcenter, or online vendor	Variable

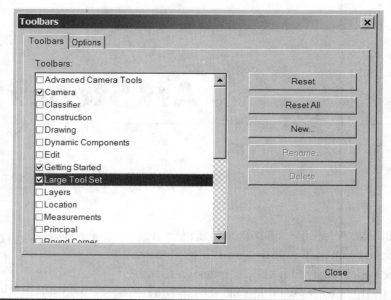

Figure 4-1   View/Toolbar shows all available tools.

Figure 4-2   The Undo tool.

arrow on the Standard toolbar (Figure 4-2) because it is a handy tool.

Draw and Offset a Circle

Click on the Circle tool (Figure 4-3), click it onto the origin (the axes' intersection), type 4", and click a second time to set the radius (Figure 4-4).

Now click on the Offset tool (Figure 4-5). Hover the mouse over the circle's perimeter—

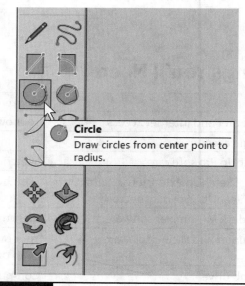

Figure 4-3   The Circle tool.

look for the On Edge popup—and drag the mouse 1/2" toward the center. You can either type that dimension and press ENTER or move the mouse until 1/2" appears in the measurements box (Figure 4-6). Click to place.

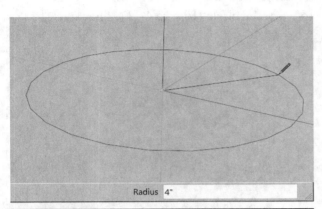

Figure 4-4   Click the circle's center, type a radius, and then click to place.

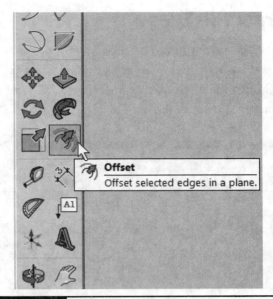

Figure 4-5   The Offset tool.

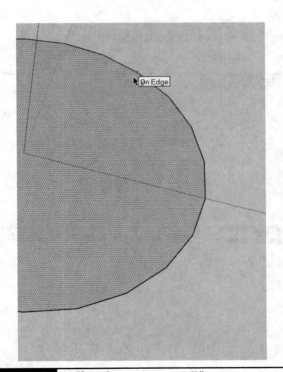

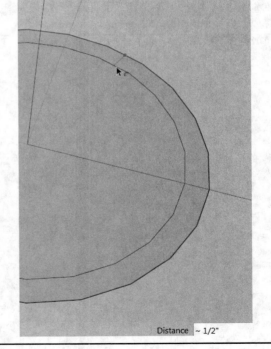

Figure 4-6   Offset the perimeter 1/2".

Draw the Eyes

Click the View toolbar's Top icon (Figure 4-7). This view makes drawing easier. Click the Circle tool onto the circle's face, and draw a 1/2"-radius circle on it (Figure 4-8).

Click the Scale tool (Figure 4-9), and then click it onto the small circle you just drew. Grips will appear (Figure 4-10). Click and drag a middle grip (it will turn red when activated) to warp the circle (Figure 4-11). Then click to finish.

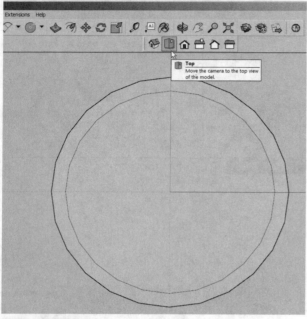

Figure 4-7   The Top icon displays the model as a top-down view.

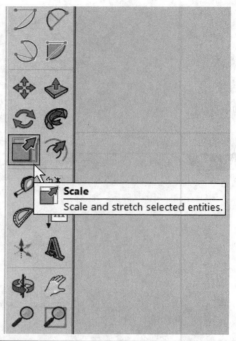

Figure 4-9   The Scale tool.

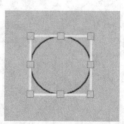

Figure 4-10   Grips appear around the circle when the Scale tool is clicked on it.

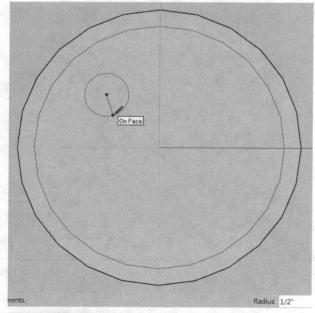

Figure 4-8   Draw a 1/2" radius circle onto the larger circle.

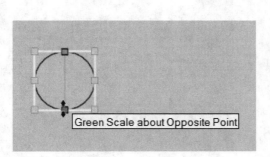

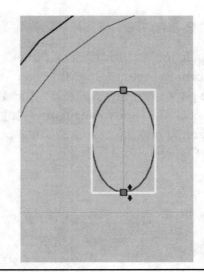

Figure 4-11　Drag a middle grip to warp the circle.

To copy the eye, click the Select tool (Figure 4-12) onto it. Click the Move tool (Figure 4-13), press the CONTROL key (OPTION key on the Mac), and move the copy off the original (Figure 4-14).

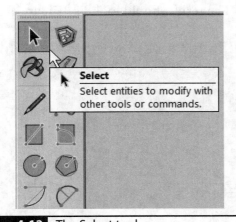

Select
Select entities to modify with other tools or commands.

Figure 4-12　The Select tool.

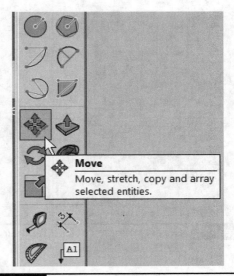

Move
Move, stretch, copy and array selected entities.

Figure 4-13　The Move tool.

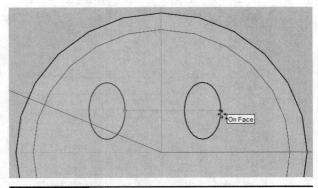

Figure 4-14　Copy the eye with Move + CONTROL key (OPTION key on the Mac).

Draw the Smile

Click the Circle tool at the origin and draw two circles as shown in Figure 4-15. Make sure that the On Face popup appears while you draw them. Then click the Line tool (Figure 4-16), and draw a horizontal line through the middle of the large circle (Figure 4-17).

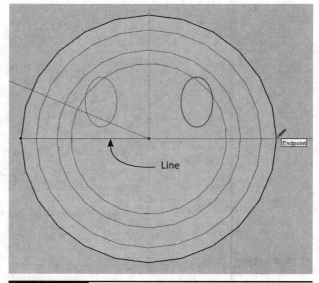

Figure 4-17 Draw a horizontal line through the center of the circle.

Draw a horizontal line at each end of the smile, as shown in Figure 4-18. Click on the Eraser tool (Figure 4-19), and click it onto the curves above the lines you just drew (Figure 4-20). The result should look like Figure 4-21.

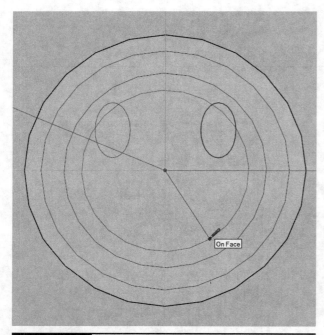

Figure 4-15 Draw two more circles with their centers at the origin.

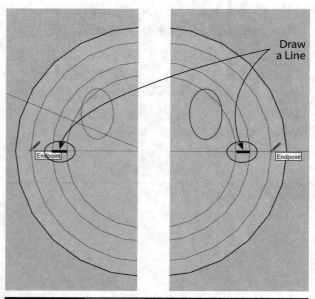

Figure 4-18 Draw horizontal lines at the smile ends.

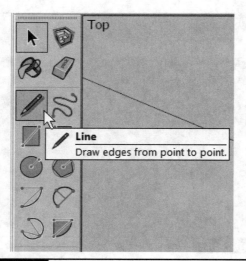

Figure 4-16 The Line tool.

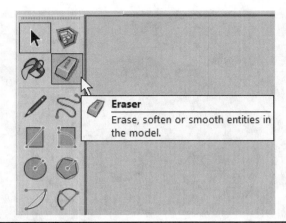

Figure 4-19 The Eraser tool.

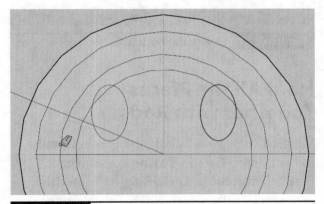

Figure4-20 Erase unneeded lines.

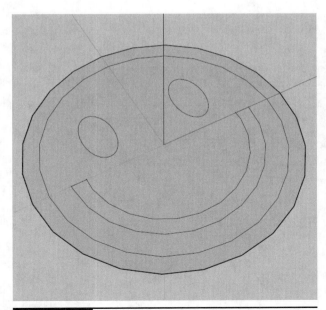

Figure 4-21 The smiley face.

Add Volume

Click on the Push/Pull tool (Figure 4-22), hover the mouse over the circle's border, and pull it up 1" (Figure 4-23). Then pull the eyes and smile up 1". Match the height of the eyes and smile to the border's height by push/pulling them up, hovering the mouse over the border, and clicking (Figure 4-24).

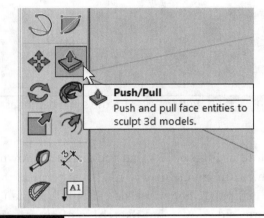

Figure 4-22 The Push/Pull tool.

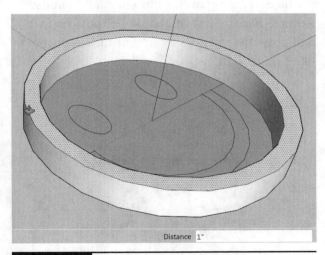

Distance 1"

Figure 4-23 Push/pull the border up.

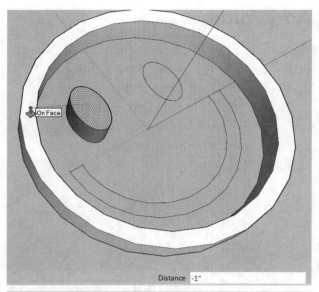

Figure 4-24 Pull the eyes up the same height as the border.

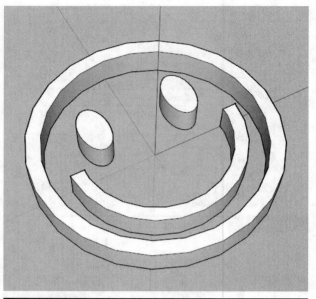

Figure 4-25 Delete the bottom face.

Note in Figure 4-24 that the distance shown in the measurements box is –1. This is because the face is flipped. The white side should be up and the gray side down, but SketchUp often gets this orientation wrong. Just click the Select tool onto the face to highlight it, right-click, and choose Reverse Faces. Next, highlight the bottom face and press the DELETE key so that the result looks like Figure 4-25.

Hold All the Pieces Together with Rods

We need some rods to hold the eyes and smile to the border. Go to File/Camera, and check Parallel Projection. This displays the model orthographically, which makes modeling the rods easier. Then click the View toolbar's Right icon (Figure 4-26).

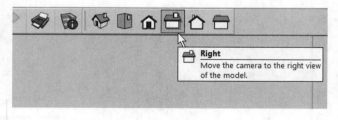

Figure 4-26 File/Camera/Parallel Projection and the Right icon display the model orthographically.

Click on the Rectangle tool (Figure 4-27), and draw a square with it by clicking two points. Then push/pull it into a box, matching its height to the smiley model (Figure 4-28). Draw a 1/8" radius circle on that box near the top and opposite the eyes (Figure 4-29). Copy the circle with the Move tool and CONTROL key (OPTION on the Mac), and move the copy opposite the smile (Figure 4-30). Then push/pull that circle through the model as shown in Figure 4-31.

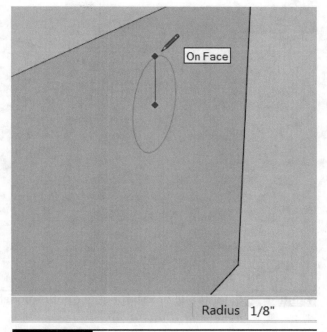

Radius 1/8"

Figure 4-29 Draw a 1/8" radius circle near the top of the box.

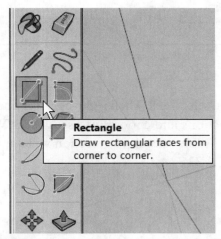

Figure 4-27 The Rectangle tool.

Rectangle
Draw rectangular faces from corner to corner.

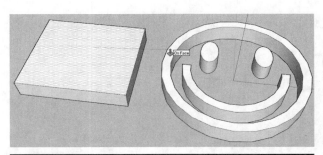

Figure 4-28 Make a box the same height as the smiley model.

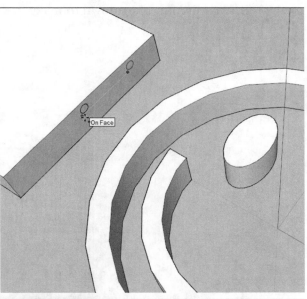

Figure 4-30 Copy the circle, making sure to move it along the green axis.

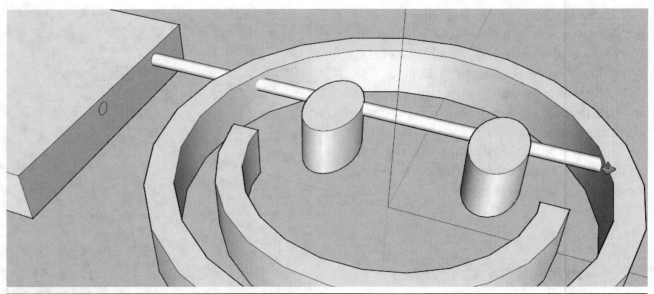

Figure 4-31   Push/pull a rod through the entire model.

Note that there's no intersection line where the rod pierces the smiley border (Figure 4-32). Fix this by double-clicking each piece of the rod; hold the SHIFT key down to make multiple selections. Then right-click on one of the selections and choose Intersect Faces/With Model. A line at each intersection will appear (Figure 4-33).

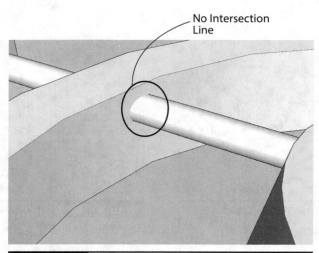

No Intersection Line

Figure 4-32   There is no intersection line where the rod pierces the model.

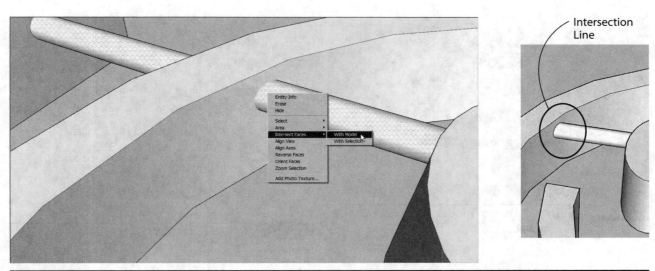

Figure 4-33 Create intersection lines with the Intersect Faces/With Model function.

Delete the Unneeded Parts

Drag the mouse around the box to select it (Figure 4-34), and then hit the DELETE key to erase it. Delete the two short rods, too. You should have a result like Figure 4-35. But let's check something before going further. Orbit underneath the model to make sure its bottom is intact. If a face is missing, trace an edge with the Line tool (Figure 4-36) to restore it (Figure 4-37).

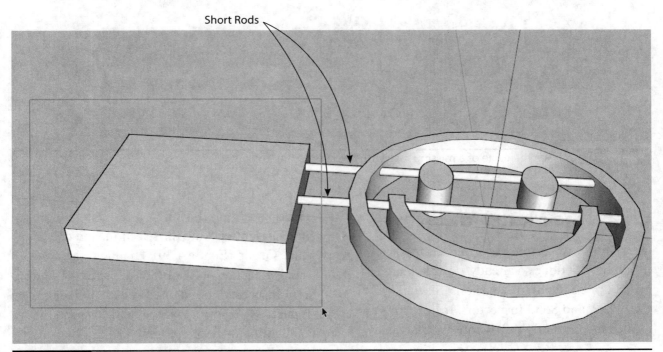

Figure 4-34 Select and delete the box and the two short rods attached to it.

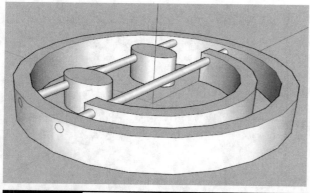

Figure 4-35 The smiley face model with extra parts removed.

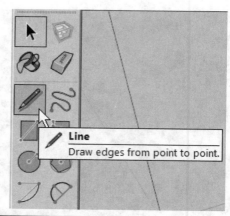

Figure 4-36 The Line tool.

Trace an Edge

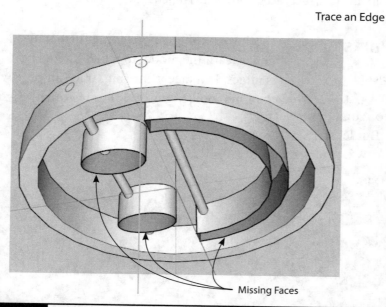

Missing Faces

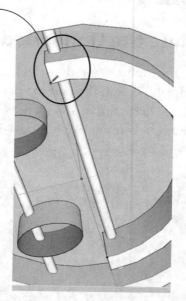

Figure 4-37 Trace the edge of a missing face with the Line tool to restore it.

Check the Model for Errors

Check for errors before exporting the model. If you haven't done so already, download and install three SketchUp extensions: SketchUp STL, CleanUp[3], and Solid Inspector[2]. Instructions are located in "General Stuff Before We Start."

Click Extensions/CleanUp[3]/Clean (Figure 4-38) to let this tool find and fix what it can. Then group the model by selecting it (either triple-

click on it or drag a window around it with the cursor), right-clicking the selection, and choosing Make Group (Figure 4-39). Highlight the group, and click on Tools/Solid Inspector[2]. Here Solid Inspector[2] found and fixed all problems (Figure 4-40). Verify that the model is solid by highlighting the group, right-clicking, choosing Entity Info, and seeing if Solid Group appears (Figure 4-41). Our smiley face shows up as a solid group, so let's export it now.

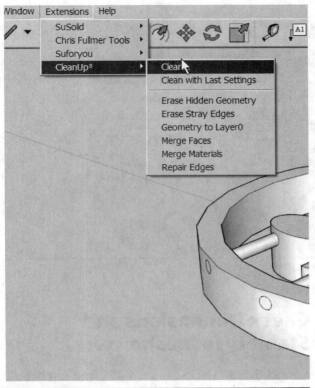

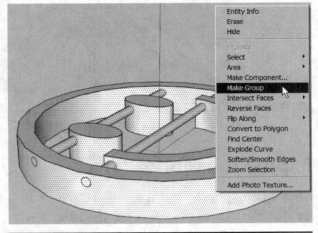

Figure 4-38   Click on Extensions/CleanUp³.

Figure 4-39   Turn the model into a group.

Figure 4-40   Click on Tools/Solid Inspector² to check the model for water-tightness.

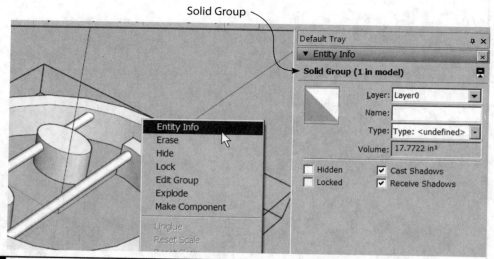

Solid Group

Figure 4-41   The Entity Info window shows that the model is a solid group.

Export the Model

Click File/Export STL (Figure 4-42), and in the window that appears, choose the model's units and Binary (it's a smaller file than ASCII).

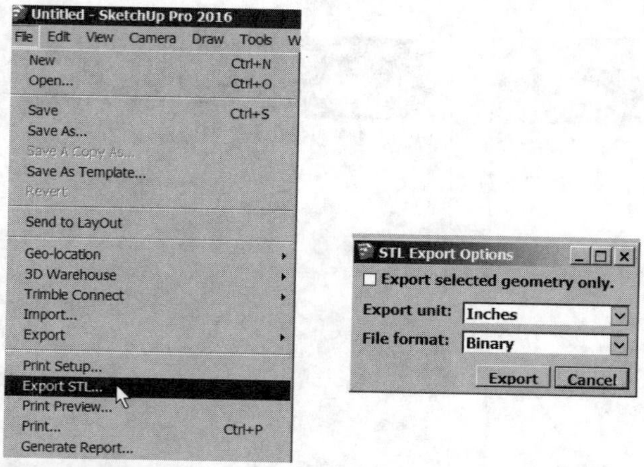

Check Dimensions and Stability in Meshmixer

Launch Meshmixer (Figure 4-43), and import the STL file. Here we'll check dimensions and stability. SketchUp STLs often export with

Figure 4-42   Export the model.

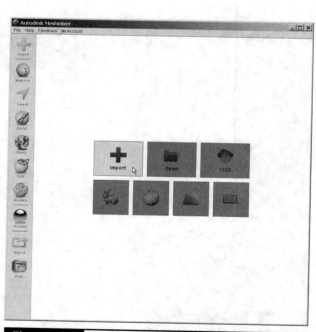

Figure 4-43   Import the STL file into Meshmixer.

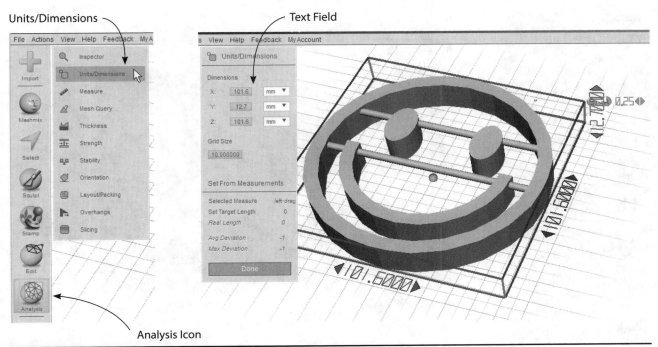

Figure 4-44 Check the model's dimensions and change if needed.

incorrect dimensions, but their proportions stay the same, so the dimensions can be easily fixed. Click on the Analysis icon and then on Units. The model's dimensions will appear (Figure 4-44). If the dimensions are incorrect, click on one of the text fields, type the correct dimension, and the others will adjust proportionately. Click *Accept*.

Now click on Analysis/Stability. This shows whether the printed model will stand up without teetering. The red circle shows the contact area (the space the model takes up on the surface), and the green ball shows that it's stable (Figure 4-45). This means that our model will lie flat on a table. Click Done.

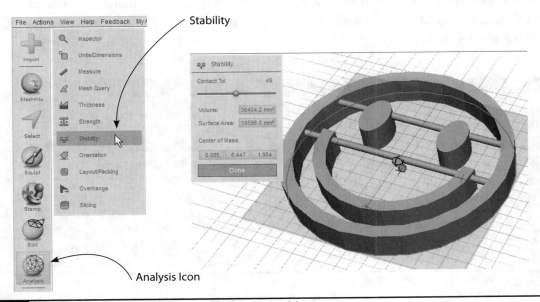

Figure 4-45 Check the model to see if it will lie flat on a table.

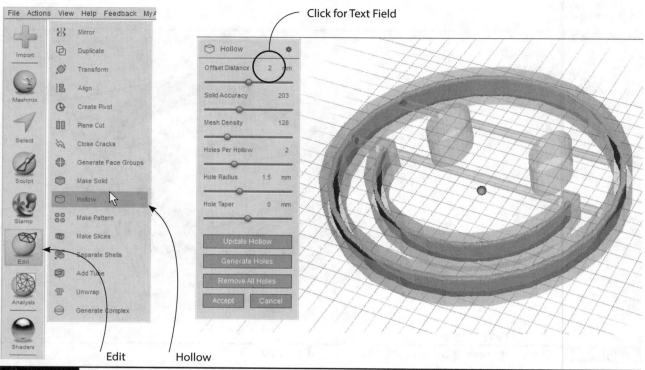

Click for Text Field

Edit   Hollow

Figure 4-46 The hollowed-out model.

Hollow the Model Out

The model will print faster and consume less plastic if it's hollowed out. Click on the Edit icon and then Hollow (Figure 4-46). The offset distance (wall thickness) is 2 mm, but you can click on it to make a text field appear and type a different distance. I typed *3* just to make the walls a bit thicker. Click Accept.

Print It!

I increased the infill to 30 percent (Figure 4-47) to make the walls stronger. The stencil was printed on a cold acrylic bed covered in painter's tape wiped with isopropyl alcohol. Figure 4-48 shows the printed stencil and some art made with it.

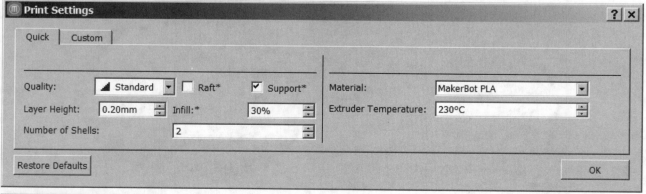

Figure 4-47 MakerBot Desktop settings.

Figure 4-48 The printed stencil and some art made with it.

Cookie Dunker

PROBLEM: A little cookie monster enjoys her bakery goods dipped in milk but doesn't like getting her fingers wet. In this project we'll make a dunker with Fusion 360, slice it with MakerBot Desktop, and print it in food-safe PLA with a MakerBot Replicator 2.

This cookie dunker consists of two parts: a holder and a handle. Launch Fusion 360, and let's start by modeling the holder.

Things You'll Need

Description	Source	Cost
Computer and Internet access	Your own or one at a makerspace	Variable
Autodesk account	Autodesk.com	Free
Autodesk Fusion 360	autodesk.com/products/fusion-360/overview	Free trial or subscription
3D printer and slicing software	Your own, one at a public makerspace, or one at an online service bureau	Variable
Thumb drive (needed only for offsite printing)	Computer or electronics store	< $10
Spool of food-safe PLA filament	Amazon, Microcenter, or online vendor	Variable
High-strength button magnets	Amazon	< $10
(Optional) Nontoxic Super Glue	Amazon	< $5

Model the Holder

This part is a box with a cutout for liquid to flow and a bump-out for gluing a magnet. We'll sketch it first and then turn it into a body (solid).

Sketch a Rectangle

On the menu at the top of the screen, click on Sketch/Rectangle/2-Point Rectangle (Figure 5-1).

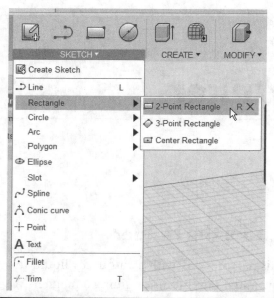

Then click two points to define the rectangle. You can hit the TAB key to toggle between text fields to type specific dimensions, but we're just going to eyeball proportions here. The file can be resized in the slicer if needed. When finished, click on Stop Sketch at the top of the screen (Figure 5-2). Return to a 3D position by either clicking the house that appears when hovering the mouse over the View Cube or by clicking the Orbit icon at the bottom of the screen (Figure 5-3). You can also orbit by holding the SHIFT key down first and then holding the mouse's scroll wheel down. Hit the ESC key to get out of Orbit mode.

Figure 5-1   Click on Sketch/Rectangle/2-Point Rectangle.

1. Click on the horizontal plane.

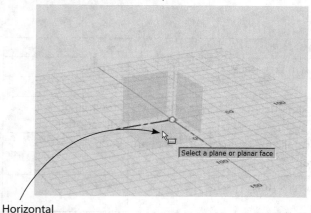

Horizontal Plane

2. Click two points.

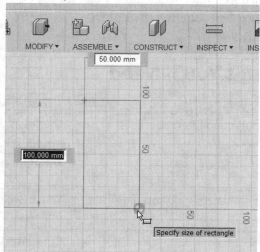

3. Click Stop Sketch.

Figure 5-2   Sketch the rectangle.

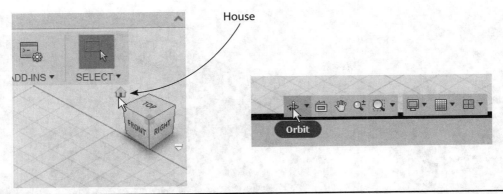

Figure 5-3 Return to a 3D view by clicking the house by the View Cube or the Orbit tool.

Extrude the Rectangle and Shell the Box

Right-click on the rectangle's face to bring up an options tree. Click on Press Pull, and extrude the rectangle up (Figure 5-4). Then right-click on the face again and choose Shell (Figure 5-5). Drag the arrow to the desired thickness (at least 2 mm to be printable), and click OK.

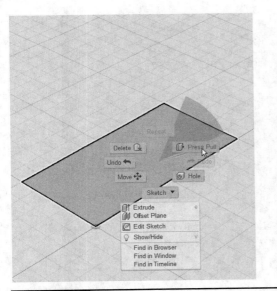

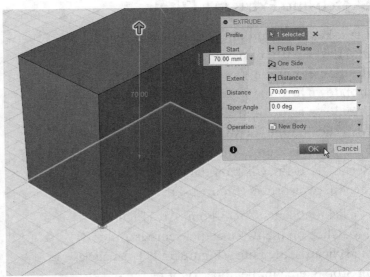

Figure 5-4 Extrude the rectangle.

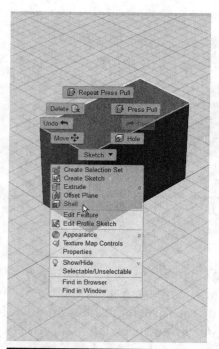

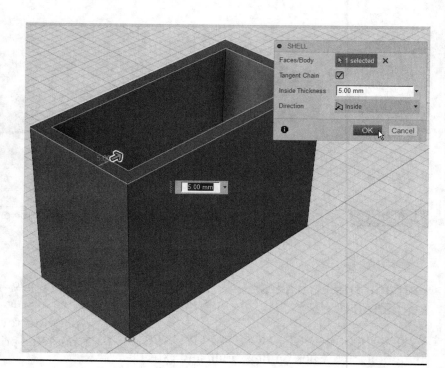

Figure 5-5 Shell the box.

Sketch and Extrude a Cutout

Let's sketch the cutout that lets liquids flow into the box. To make sketching easier, hover the mouse over the View Cube to make the dropdown arrow appear, then click on the dropdown arrow, and click on Orthographic. Also click on the View Cube's right side (Figure 5-6). Now click on Sketch/Rectangle/2-Point Rectangle again. Click on the vertical plane and then on two corners (Figure 5-7).

Click on Sketch/Arc/3-Point Arc (Figure 5-8). Click on the two rectangle corners, and then click the arc bulge (Figure 5-9). Then click on Stop Sketch in the upper right of the screen (Figure 5-10).

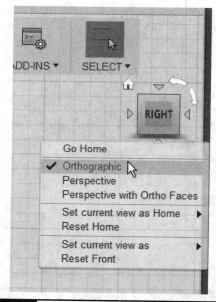

Figure 5-6 View the work plane orthographically.

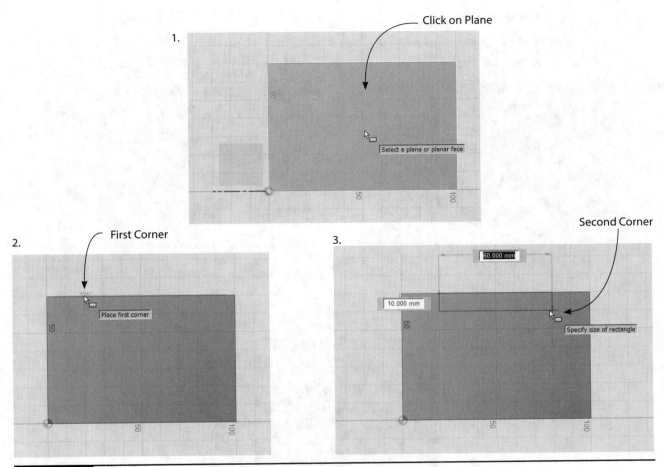

Figure 5-7   Sketch a rectangle.

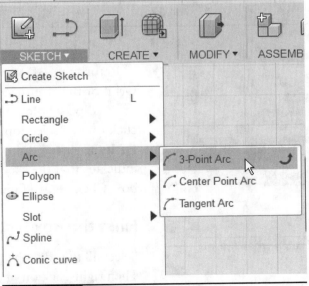

Figure 5-8   Click on Sketch/Arc/3-Point Arc.

1. First Point

2. Second Point

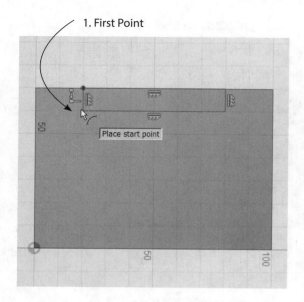

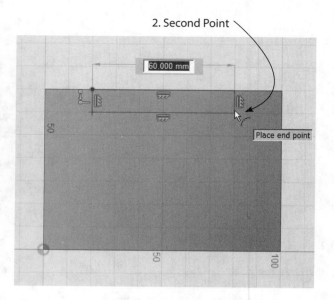

3. Arc Bulge

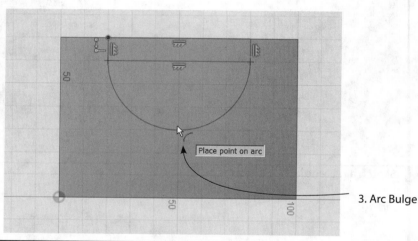

Figure 5-9   Sketch an arc.

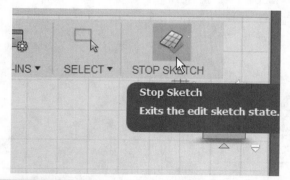

Figure 5-10   Click on Stop Sketch when finished sketching.

Return to a 3D view by clicking either the Orbit or the House icon. Then select both parts of the dip by holding the SHIFT key down. Right-click on one selection and choose Extrude. Push the sketch through the box. The red color indicates that the sketch is cutting through the box. Click OK to finish (Figure 5-11).

Fillet the Box

Select all the edges by holding the SHIFT key down. Then right-click on one edge and choose Fillet. Drag the arrow or type a fillet dimension (Figure 5-12) to round the edges off. Click OK to finish.

1.

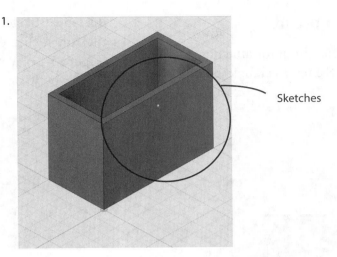

Sketches

2.

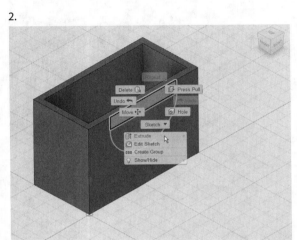

3.

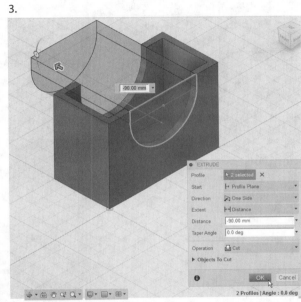

Figure 5-11 Extrude the rectangle and arc sketches through the box.

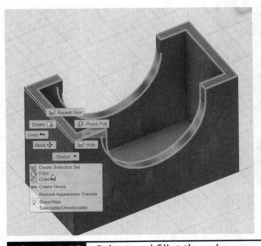

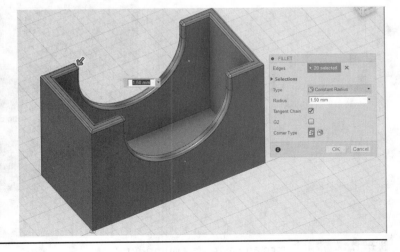

Figure 5-12 Select and fillet the edges.

Model the Bump-Out

The box needs a bump-out for attaching a magnet. Click on Sketch/Rectangle/2-Point Rectangle. Sketch a rectangle on the box, right-click on it, click on Press Pull, and extrude it (Figure 5-13). You can select and fillet its edges, too, if you wish.

1.

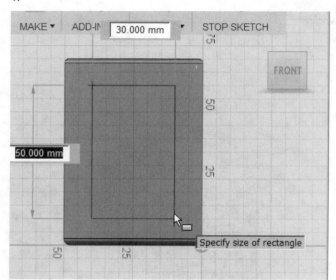

2.

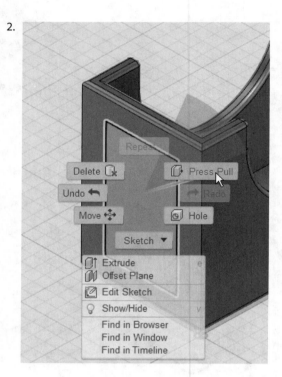

3.

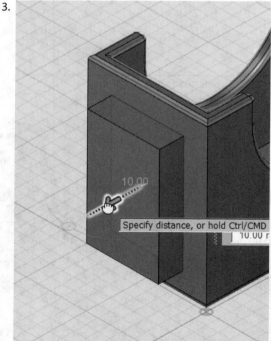

Figure 5-13 Sketch and extrude a bump-out.

Model the Handle

The handle will be a box and cylinder to which a magnet can be glued. The box needs to be the same size as the bump-out. Simply copying and pasting the original sketch rectangle to extrude another box isn't the easy operation in parametric software (which Fusion 360 is) that it is in nonparametric software, so we're going to use a different technique.

Offset a Plane

In the Browser window, click the dropdown arrow in front of Sketches. When you click an entry there, the corresponding sketch in the model will become selected. Find the sketch rectangle listing, right-click, and choose Offset Plane. Then drag the plane forward (Figure 5-14).

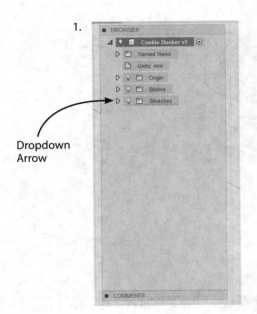

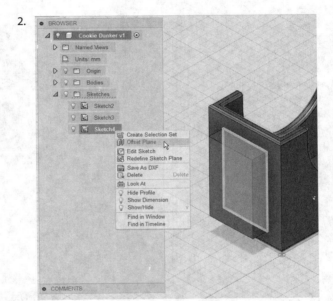

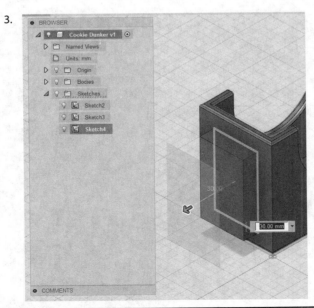

Figure 5-14 Offset the rectangle sketch plane.

Sketch a Rectangle around the Plane

Next, click on the Rectangle tool. The plane disappears after you do this, so click the dropdown arrow in front of Construction, and click on Plane 1 to make the light bulb in front of this listing yellow (Figure 5-15), which will make the plane reappear. When the light bulb in front of a listing is gray, that item is invisible. Sketch a rectangle over the plane (Figure 5-16). Then right-click it, click on Press Pull, and extrude it forward (Figure 5-17).

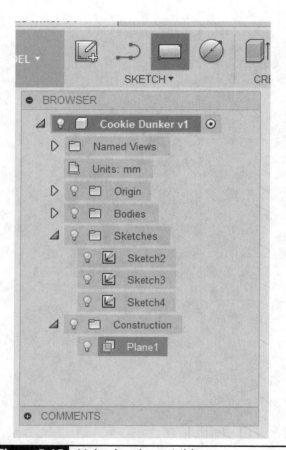

Figure 5-15 Make the plane visible.

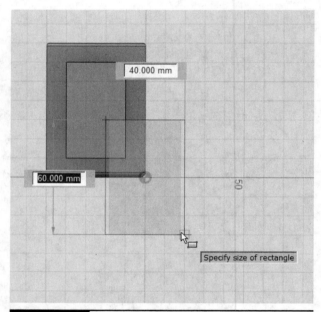

Figure 5-16 Sketch a rectangle around the plane.

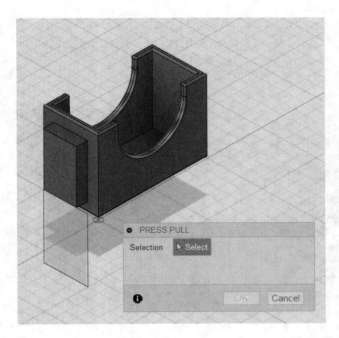
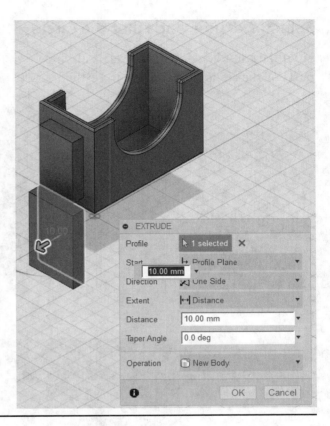

Figure 5-17 Press/pull the plane.

Sketch a Circle on the Handle

Click on Sketch/Circle/Center Diameter Circle (Figure 5-18). Click it onto the center of the new box (it helps if you align the box with the grid so that you can use the grid to find the center). Click a diameter that is larger than the width of the bump-out.

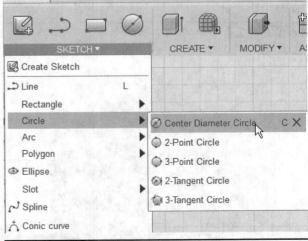

Figure 5-18 Click on Sketch/Circle/Center Diameter Circle.

Select, Extrude, and Fillet the Circle

Hold the SHIFT key down, and select all three parts of the circle (it is broken up by the rectangle). Right-click, click on Press Pull, and extrude it (Figure 5-19). Then select its edges and fillet (Figure 5-20).

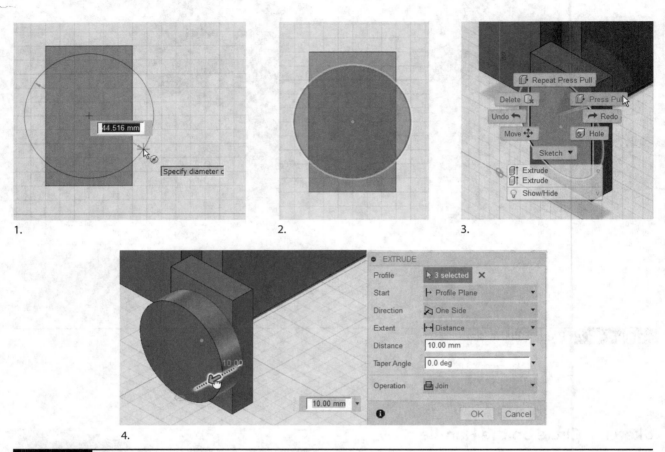

1.

2.

3.

4.

Figure 5-19   Sketch a circle, select it, and extrude it.

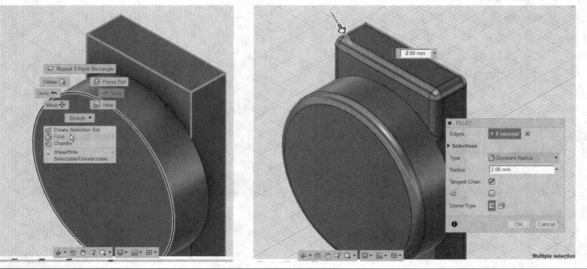

Figure 5-20   Fillet the circle.

Export the Dunker and Handle Separately

The individual parts (box, bump-out, circle on handle) don't need to be combined because they fused together when the sketches on them were extruded. We just need to export the dunker and handle as separate STL files. First, right-click on the handle, click Move, and drag the button to rotate it as shown in Figure 5-21. This orientation is easier to print. Right-click on it again, and click Show/Hide to hide it (Figure 5-22). Then right-click on the dunker file's name

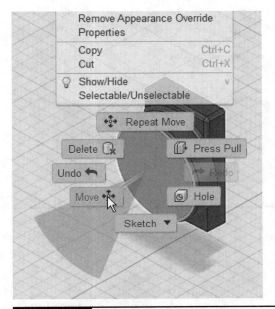

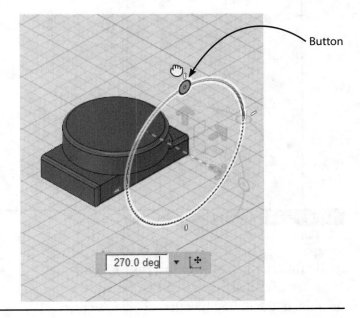

Figure 5-21   Rotate the handle.

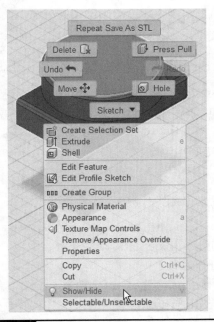

Figure 5-22   Hide the handle before exporting the dunker.

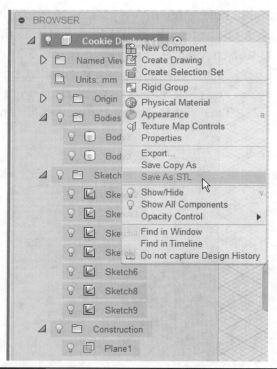

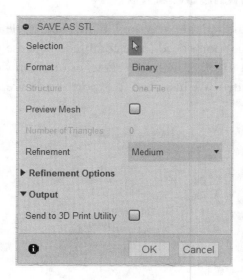

Figure 5-23 Export the dunker as an STL file.

in the Browser window (in this case, Cookie Dunker), and click Save As STL. Accept the defaults in the dialog box that appear (Figure 5-23). Next, hide the dunker (Figure 5-24), and export the handle as an STL file.

Print It!

Figure 5-25 shows the cookie dunker positioned inside MakerBot Desktop. You can move and rotate it inside MakerBot Desktop with the red Rotation button for a better viewing position during printing. I set the infill to 10 percent to make the dunker light enough for button magnets to support it. Stronger magnets than typical craft buttons are needed; you can find some at Amazon. Attach them with nontoxic super glue. Figures 5-26 and 5-27 shows the finished cookie dunker. The cutter was printed on a cold acrylic bed covered with painter's tape wiped with isopropyl alcohol.

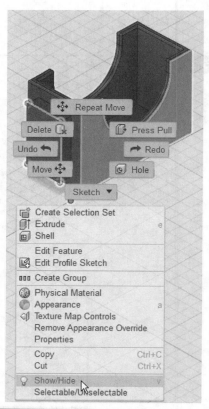

Figure 5-24 Hide the dunker before exporting the handle.

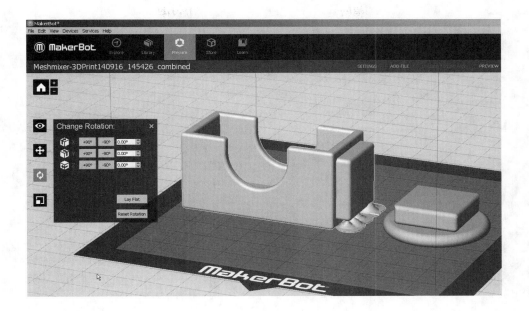

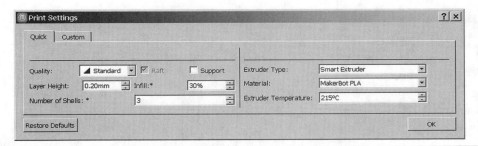

Figure 5-25   The position and settings in MakerBot Desktop.

Figure 5-26   The cookie dunker with attached magnets.

Figure 5-27   The cookie dunker in use.

Simple Bat-Shaped Cookie Cutter

PROBLEM: A parent needs a bat-shaped cutter to make cookies for her son's school Halloween party. In this project we'll use a downloaded PNG file, an online file converter, Fusion 360, and Meshmixer to make a cutter. We'll slice it with Simplify3D and print it in one color with PLA on a Lulzbot Taz 6.

This cutter will be a simple outline. It can be modeled from a sketch you manually or digitally draw yourself or from one you find online and convert to an appropriate file format. We'll do the latter.

Things You'll Need

Description	Source	Cost
Computer and Internet access	Your own or one at a makerspace	Variable
Google account	accounts.google.com	Free
Autodesk account	Autodesk.com	Free
Autodesk Fusion 360	autodesk.com/products/fusion-360/overview	Free trial or subscription
Autodesk Meshmixer software	Meshmixer.com	Free
3D printer and slicing software	Your own, one at a public makerspace, or one at an online service bureau	Variable
Thumb drive (needed only for offsite printing)	Computer or electronics store	< $10
Spool of PLA filament	Amazon, Microcenter, or online vendor	Variable

Figure 6-1   Results from a Google image search for bat sketches.

Find and Convert a Bat Sketch

Finding an existing sketch to model is often quicker than drawing one from scratch. Do a Google image search for bat sketches (Figure 6-1). A simple, crisp black and white image is best because colors and details often convert poorly. Figure 6-2 shows a PNG image that I found.

Point your browser to image.online-convert .com, and click *Convert image to the SVG format* (Figure 6-3). Then navigate to the bat PNG file,

Figure 6-2   The PNG file chosen.

Figure 6-3   At image.online-convert.com, click Convert image to the SVG format.

Upload
File Here

Convert File
Button

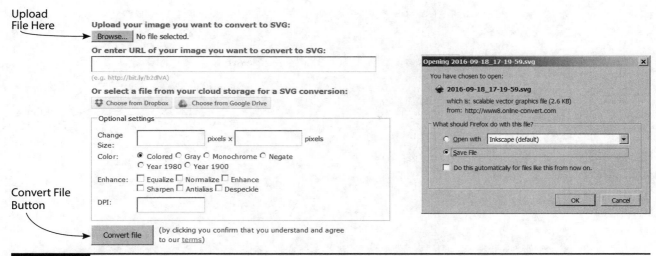

Figure 6-4 Upload the file, click the Convert file button, and then download the SVG file.

upload it, keep the default settings, click the *Convert file* button (Figure 6-4), and download the SVG file. Then launch Fusion 360.

Import the SVG File into Fusion 360

On the menu at the top of the screen, click Insert/Insert SVG (Figure 6-5). A dialog box and three planes will appear (Figure 6-6). Click on the file folder, and navigate to the SVG file. Then click on the horizontal plane. The file will enter (Figure 6-7). You can drag the arrows to move it and change its size by typing new dimensions in the text field. This file entered with 0 as the dimension in the *x* and *y* field, so I typed *76* in the *x* field, the size I want the cutter to be. If you leave the default dimension at 0, the file may disappear after you click OK.

Figure 6-5 Click Insert/Insert SVG.

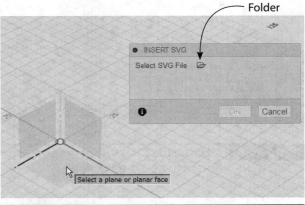

Figure 6-6 Click on the folder to navigate to the SVG file, and then click on the horizontal plane.

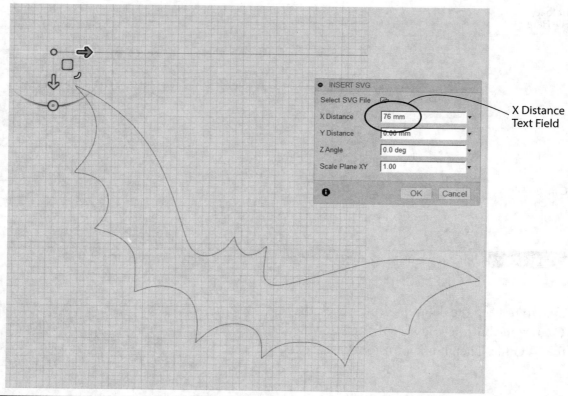

INSERT SVG

Select SVG File

X Distance   76 mm   ⟵ X Distance Text Field

Y Distance   0.00 mm

Z Angle   0.0 deg

Scale Plane XY   1.00

OK   Cancel

Figure 6-7   The imported SVG file with its *x* dimension set to 76.

Return the workspace to a 3D display by either clicking on the Orbit tool or hovering the mouse over the View Cube to make the house appear and then clicking on the house (Figure 6-8).

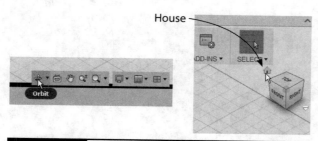

House

Orbit

Figure 6-8   Click on the Orbit tool or the house to return to a 3D display.

Offset the Sketch

Click Sketch/Offset (Figure 6-9). Click anywhere on the sketch. It should highlight as one piece. Either drag the offset with the arrow or type a dimension in the text field (Figure 6-10). I typed *1 mm*.

Tip: Some SVG sketches will import as dozens of little pieces, making offsetting the whole sketch tedious or impossible. You never know if this will be the case with a Web file. A workaround is to copy the sketch, scale the copy down, and move the smaller copy onto the larger one. Copy it by selecting, pressing CTRL + C and CTRL + V, then dragging the copy off the original with the arrows that appear.

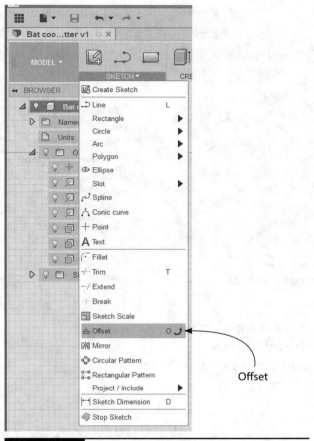

Figure 6-9 Click Sketch/Offset.

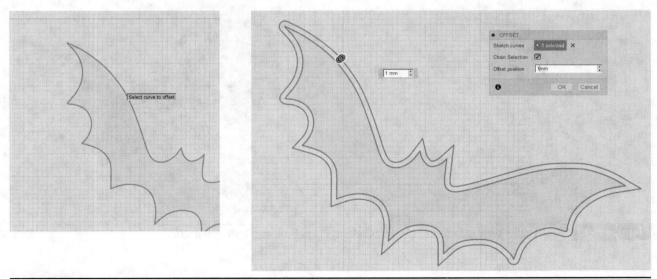

Figure 6-10 Offset the sketch.

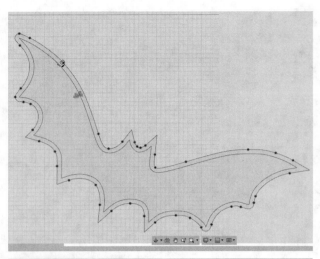

Figure 6-11 Inflection points appear that can be dragged to change the shape.

A bunch of inflection points will appear on the offset (Figure 6-11). You can drag these points to adjust the shape, but we're going to leave them alone. Right-click on the face to select it, choose Press Pull, and extrude it up (Figure 6-12). I extruded it up 12 mm.

> ***Tip:*** If you decide that the cutter's size isn't right, you can change it. To find out its current size, click on Inspect/Measure, click on the first icon in the Measure dialog box, and then click on opposite ends of the cutter to find the distance between them. To change its current size, click Modify/Scale and either drag the arrow to eyeball a new size or click a specific scale factor into the dialog box. To scale it to a specific size, divide the size you want by the size you have. You can also scale the cutter nonuniformly by clicking on the dropdown arrow next to Uniform in the Scale dialog box.

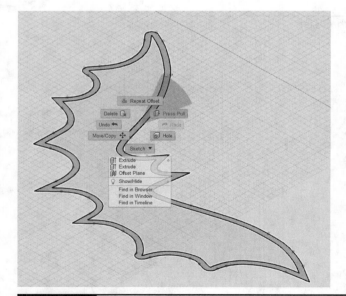

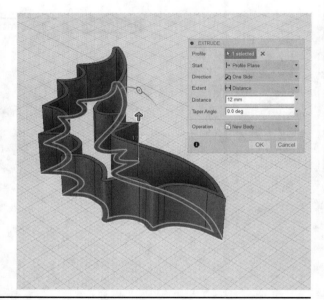

Figure 6-12 Right-click on the face to select it, click on Press Pull, and extrude it up.

Export the Cutter as an STL File

Right-click on the file's name in the Browser window (in this case, Bat cookie cutter). Click on Save as STL, accept the defaults in the dialog box, and click OK (Figure 6-13).

Check for Stability

Let's check if this cutter will sit flat on a surface before printing it. Import the STL file into Meshmixer (Figure 6-14). Files often import into this program at odd angles. To fix, hit the T key on the keyboard and reorient with the arrows

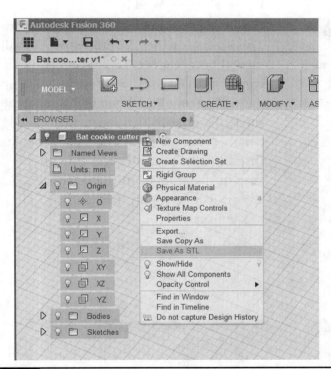

Figure 6-13 Right-click on the file's name, and click on Save as STL.

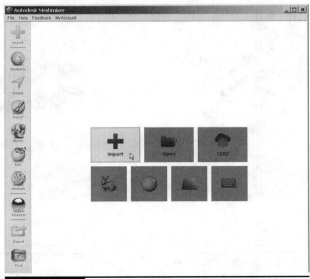

Figure 6-14 Import the STL file into Meshmixer.

and curves on the manipulator that appears (Figure 6-15). Dragging the mouse directly over the hashmarks moves the file in whole-number increments.

In this case, reorienting is easier by simply clicking Analysis/Orientation. Meshmixer will orient the file in an optimal manner for 3D printing. Click Accept.

Now click on Analysis/Stability (Figure 6-16). This shows whether the printed model will stand up without teetering. The red circle shows the contact area (the space the model takes up on the surface), and the green ball shows that it's stable. This means that our model will lie flat on a table. Click Done. You can also click Analysis/Inspect if you want to check for defects that will keep the file from being 3D printable. Click the Export icon to export the file as an STL file from Meshmixer.

Figure 6-15   Hit the т key to bring up a transform tool for reorienting the file.

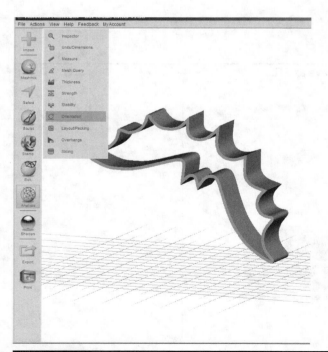

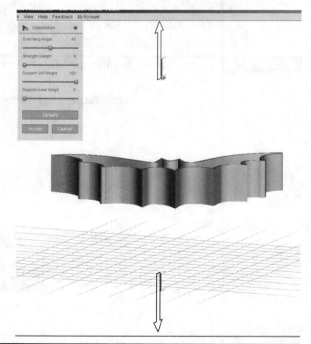

Figure 6-16   Reorient the model if necessary at Analysis/Orientation.

Try to Print It!

I sliced the file in Simplify3D and used a raft (Fig 6-17). The first attempt at printing failed because the Taz couldn't make the sharp edges of the ears and wings on the 2-mm-wide walls. It deposited string filament and then "air printed." I doubled the size but got the same result.

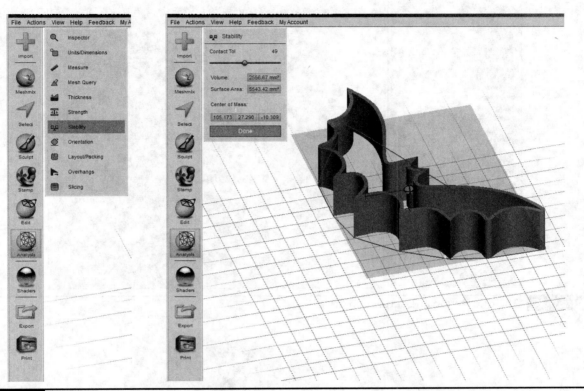

Figure 6-17 Check for stability at Analysis/Stability.

Print It!

Then I sliced it in Cura, also doubling the size. Cura sliced it differently, and the result was successful. Figure 6-18 shows the printed cutter, and Figure 6-19 shows some cut dough. The cutter was printed on a hot build plate covered with PEI and wiped with isopropyl alcohol.

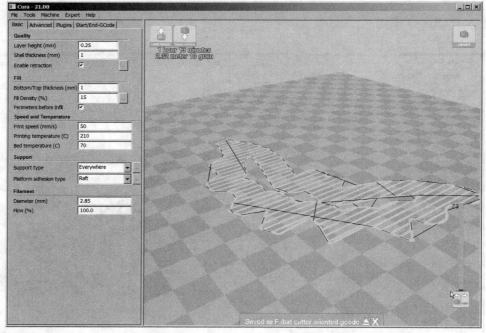

Figure 6-18   The orientation and settings in Cura.

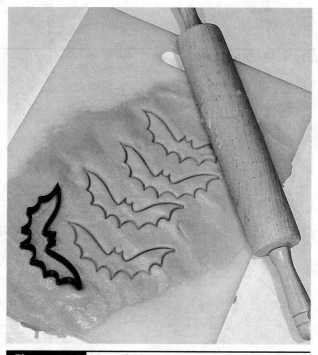

Figure 6-19   Dough cutouts.

Spiral Ornament

PROBLEM: An Etsy seller wants a spiral design for a holiday ornament. We'll draw it in Inkscape, model it in SketchUp Pro, size it in Meshmixer, slice it in Simplify3D, and print it in PLA on a MakerBot Mini.

Draw the Spiral

Launch Inkscape. Click on the Spiral tool (Figure 7-1). Then click it inside the workspace, but don't release the mouse—instead, drag

Things You'll Need

Description	Source	Cost
Computer and Internet access	Your own or one at a makerspace	Variable
Google account	accounts.google.com	Free
SketchUp Pro software	Sketchup.com	Free trial or $690
Three SketchUp extensions: SketchUp STL, CleanUp[3], and Solid Inspector[2]	SketchUp Extension Warehouse	Free
3D printer and slicing software	Your own, one at a public makerspace, or one at an online service bureau	Variable
Thumb drive (needed only for offsite printing)	Computer or electronics store	< $10
Spool of PLA filament	Amazon, Microcenter, or online vendor	Variable

it. The spiral won't appear in the workspace (Figure 7-2) until you start dragging it. Release the mouse when the spiral is the size you want (Figure 7-3). Then click on File/Save As, and save it as a DXF file (Figure 7-4).

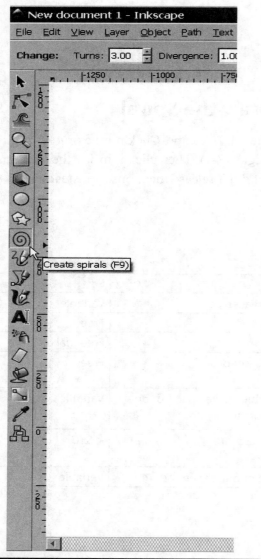

Figure 7-1 The Spiral tool.

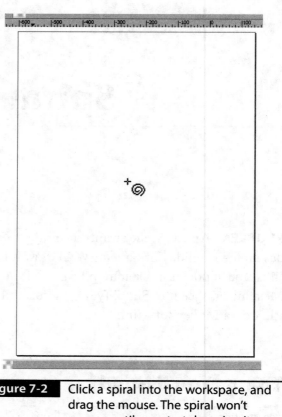

Figure 7-2 Click a spiral into the workspace, and drag the mouse. The spiral won't appear until you start dragging it.

Figure 7-3 Release the mouse when the spiral is the size you want.

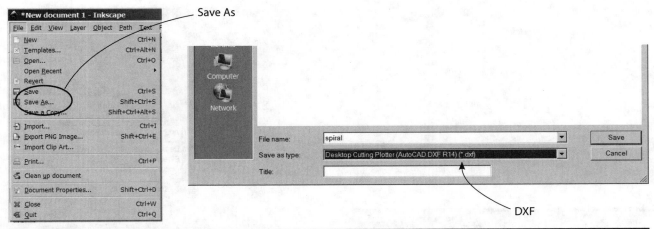

Figure 7-4 Save the spiral as a DXF file.

Import the Spiral into SketchUp Pro

Launch SketchUp Pro, click on File/Import, and set the text field at the bottom of the navigation window to AutoCAD files. Navigate to the

spiral DXF, and click Import (Figure 7-5). The imported file may be very small, so click on the Zoom Extents tool to find it (Figure 7-6). If you see endpoints all over it, turn them off by going to the Default Tray on the right side of the workspace, clicking on Style/Edit, and

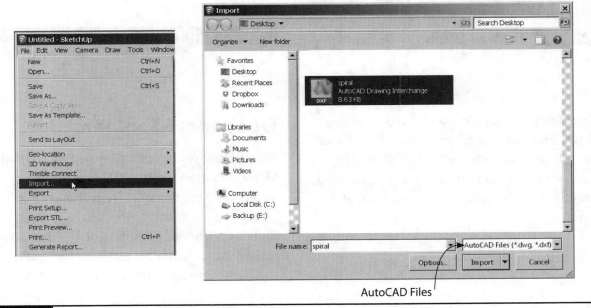

Figure 7-5 Import the Spiral DXF file into SketchUp.

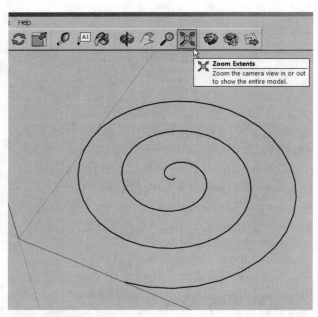

Figure 7-6 | The Zoom Extents tool fills the screen with the model.

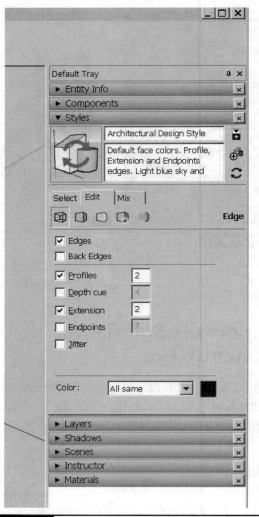

Figure 7-7 | If the endpoints are visible, uncheck the Endpoints box at Style/Edit.

unchecking the Endpoints box (Figure 7-7). This will make the model smoother. And if you haven't done so already, install the SketchUp STL, Solid Inspector[2], and CleanUp[3] extensions. Instructions are located in "General Stuff Before We Start."

Open More Toolbars

SketchUp's workspace opens with the default Getting Started toolbar. Click on Views/ Toolbars to see more toolbars (Figure 7-8). Check the boxes in front of Large Toolset, Standard, Views, and Solid Tools. Then close the screen. Note the Undo arrow on the Standard toolbar (Figure 7-9) because it is a handy tool.

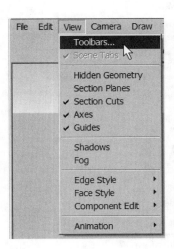
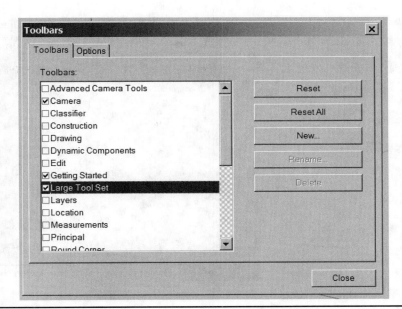

Figure 7-8 Click on View/Toolbars to see more toolbars.

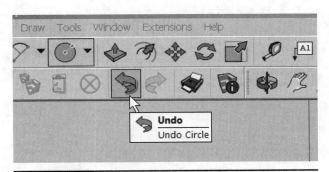

Figure 7-9 The Undo arrow on the Standard toolbar.

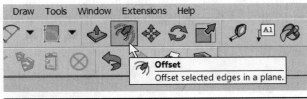

Figure 7-10 The Offset tool.

Offset the Spiral

Select the whole spiral by dragging a window around it with the mouse. Click the Offset tool (Figure 7-10), and click it onto the spiral (Figure 7-11). Drag the offset spiral outside the original spiral until it's the width you want (Figure 7-12), and click to finish.

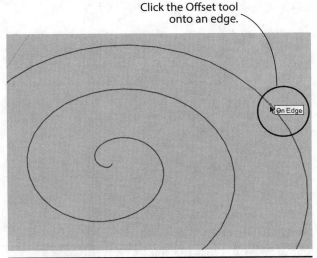

Click the Offset tool onto an edge.

Figure 7-11 Click the Offset tool onto the spiral. Look for the On Edge popup.

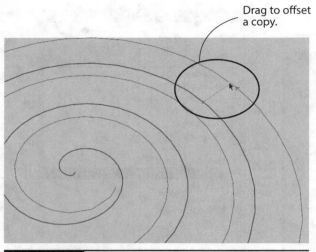

Drag to offset
a copy.

Figure 7-12   Offset the spiral to the width desired.

Click on the Line tool (Figure 7-13), and close
the spiral's ends by drawing a line between the
endpoints. A face will form (Figure 7-14).

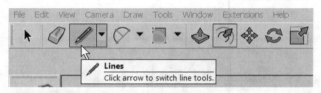

Figure 7-13   The Line tool.

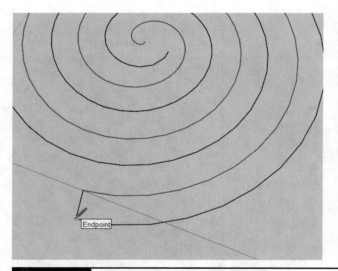

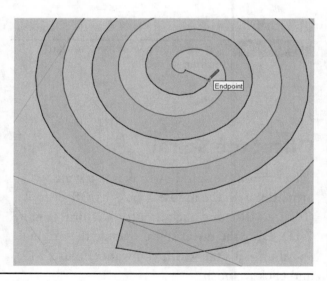

Figure 7-14   Close the spiral's ends to create a face.

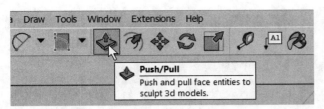

Figure 7-15 The Push/Pull tool.

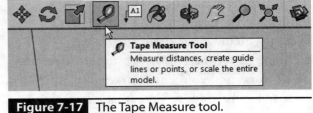

Figure 7-17 The Tape Measure tool.

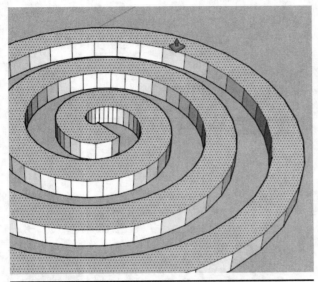

Figure 7-16 Pull the face up to the height desired.

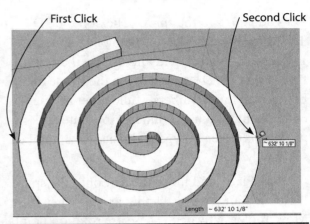

Figure 7-18 Click the Tape Measure at opposite ends of the spiral to see its current size.

Click on the Push/Pull tool (Figure 7-15), hover it over the spiral, drag it up to the height you want, and click to place (Figure 7-16).

Scale the Spiral

Click on the Tape Measure tool (Figure 7-17). Then click it on opposite ends of the spiral. Its current size will appear in a popup and in the Measurements box (Figure 7-18). We see that it's 632' wide. Type *4* to make it 4" between the clicked points (you can type anywhere; you don't have to type inside the Measurements box). A dialog box appears asking if you want to resize the model (Figure 7-19). Click Yes, and the model adjusts to the new size. Because it is now so much smaller, click the Zoom Extents tool to find it.

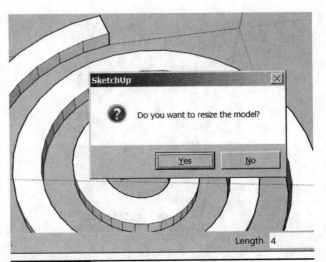

Figure 7-19 Type *4* to change that dimension to 4" wide.

Typing the new dimension must be done immediately after the second Tape Measure click; if you move the mouse, orbit, or do anything in between these two tasks, the resizing won't work, and you'll need to click the spiral's opposite ends with the Tape Measure again.

Add Text to the Spiral

Let's add *2016* to the spiral. Click on the 3D Text tool (Figure 7-20). A dialog box appears; I typed *2* in the text field and *0.25* in the Height field (Figure 7-21). I didn't type the whole date because all four numbers would enter as one component, making it difficult to rotate and place. It is easier to type and place each number separately.

Click Place, and move the text onto the spiral. Note the four red crosses that appear when the Move tool is activated. These are rotation points. Click on a point, move the mouse to position the number, and click again to set the new position (Figure 7-22).

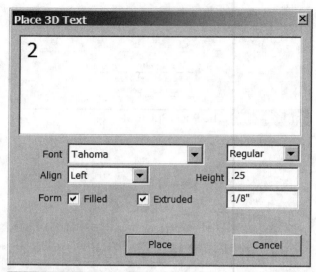

Figure 7-21   Type the text you want into this dialog box.

Figure 7-22   Position the number by rotating it around one of the four red crosses.

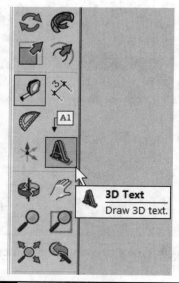

Figure 7-20   The 3D Text tool.

To place the number a specific distance from the bottom edge, click the Tape Measure tool onto the bottom edge, and drag it up a bit. This creates a guideline. Then move and click the number onto that guideline (Figure 7-23). Repeat these steps to create and place the rest of the numbers (Figure 7-24). Remove guidelines when finished by selecting and deleting them or by clicking on Edit/Delete Guides.

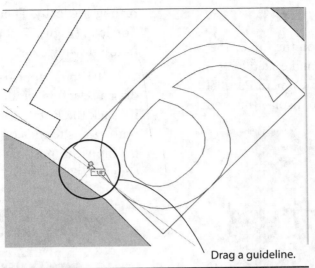

Drag a guideline.

Figure 7-23 Drag a guideline, and click the number onto it.

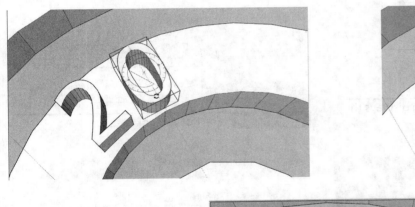

Figure 7-24 Model and place each number separately.

Reposition the Spiral

We need a neck and loop on the top of the spiral. Let's position it to make sketching them easier. First, click on Camera/Parallel Projection (Figure 7-25) to display the model

orthographically. Then click on the View menu's Top icon (Figure 7-26) to see the model in plan (a top-down view).

If the spiral appears sideways or upside-down, drag a selection window around it to highlight it. Click the Rotate tool (Figure 7-27), click it onto opposite ends of the spiral, and then drag the mouse to position the spiral right-side up (Figure 7-28).

Figure 7-25   Camera/Parallel Projection displays the model orthographically.

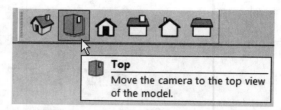

Figure 7-26   The View menu's Top icon displays a plan view of the model.

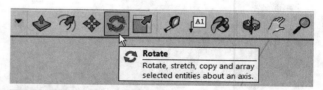

Figure 7-27   The Rotate tool.

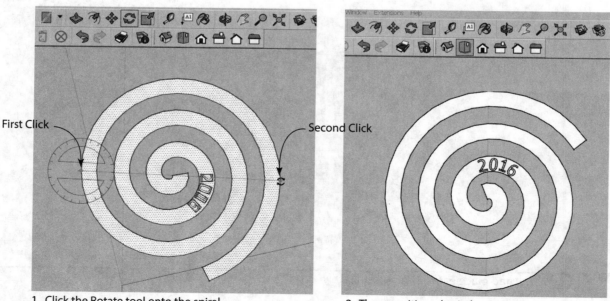

First Click

Second Click

1. Click the Rotate tool onto the spiral.

2. The repositioned spiral.

Figure 7-28   Rotate the spiral upright.

Add a Neck and Loop

To model the neck, click on the Line tool and draw a rectangle as shown in Figure 7-29. Note its alignment with the green axis. Then click the View menu's Iso icon (Figure 7-30) to return to a 3D view. Push/pull the rectangle down, clicking it on the spiral's bottom edge to match its height (Figure 7-31). That's the neck.

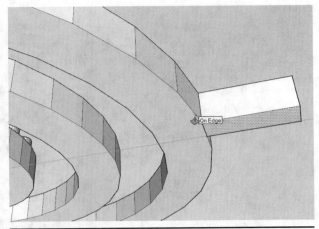

Figure 7-31   Push/pull the rectangle down.

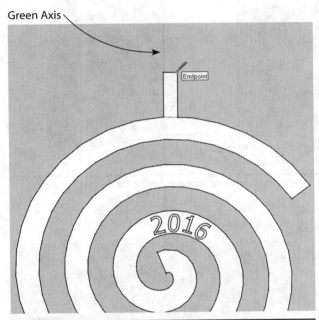

Figure 7-29   Draw a rectangle that is aligned with the green axis.

To make the loop, use the Circle tool (Figure 7-32). First, drag one of the neck's vertical edges with the Tape Measure to make a center guideline. Then click the Circle tool onto that guideline, and draw two concentric circles to make a ring (Figure 7-33). Push/pull the ring down, and click on the neck to match its depth. Select and delete the center (Figure 7-34). Finally, run the CleanUp³ extension (click on Extensions/CleanUp³/Clean) to make the model as tidy as possible.

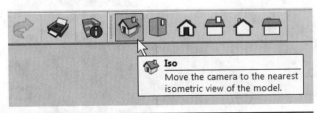

Figure 7-30   The Iso icon.

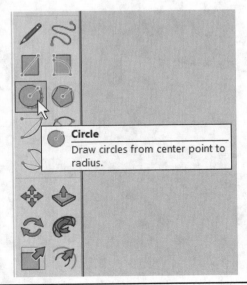

Figure 7-32   The Circle tool.

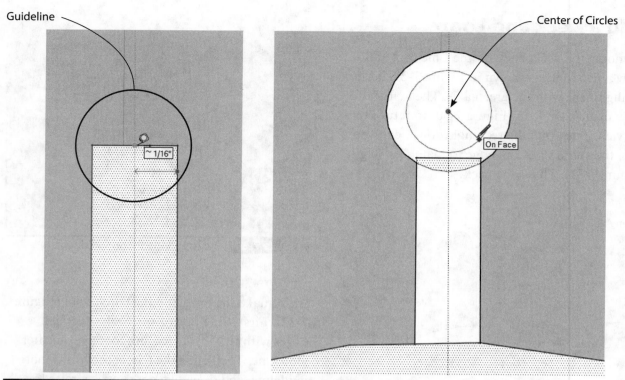

Figure 7-33 Place a guideline at the center of the rectangle, and then draw two concentric circles with their centers on the guideline.

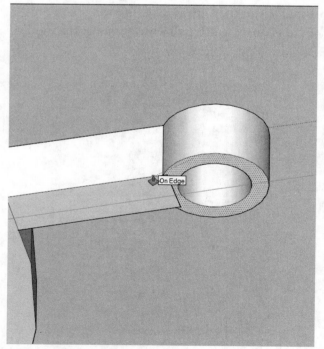

1. Push/pull the ring down and click it on the neck.

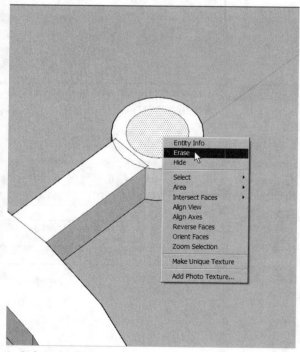

2. Delete the face.

Figure 7-34 Model the loop.

Combine All the Parts

The ornament must be one solid piece to be 3D printable. Triple-click on the spiral to select it—don't drag a selection window around it because that will select the component numbers, too. Then right-click and choose Make Group (Figure 7-35). This will group the spiral, neck, and loop together.

Click Tools/Solid Inspector[2] onto the group to find and fix any problems. Then select and right-click on the group and choose Entity Info. If the group is solid, it will say so (Figure 7-36). This group is solid, which enables us to use

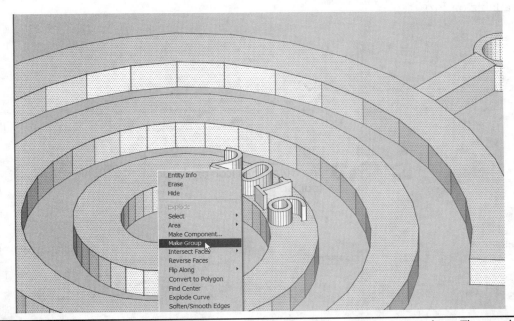

Figure 7-35   Triple-click on the spiral to select everything except the component numbers. Then make it a group.

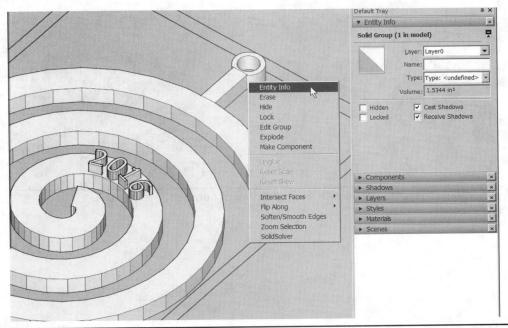

Figure 7-36   The spiral, neck, and loop are one solid group. Note that the numbers aren't included in this group. They're separate components.

Solid Tools to combine it with the component numbers (which are also solids).

Click on the Outer Shell icon on the Solid toolbar (Figure 7-37). This tool combines all groups and components and removes anything inside them. It's available on both SketchUp Make and Pro; the rest of the Solid tools icons are only available on Pro.

Click on the spiral group. Then click on each number (Figure 7-38) to combine them into one piece. Verify that the ornament is one piece by clicking anywhere on it to see if it all selects (Figure 7-39). Then click File/Export STL (Figure 7-40).

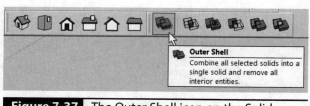

Figure 7-37 The Outer Shell icon on the Solid toolbar.

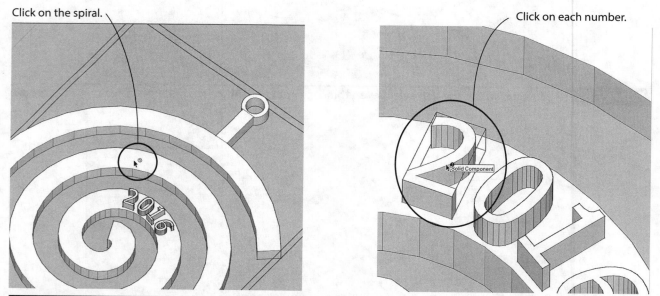

Figure 7-38 Combine the spiral group with each component number.

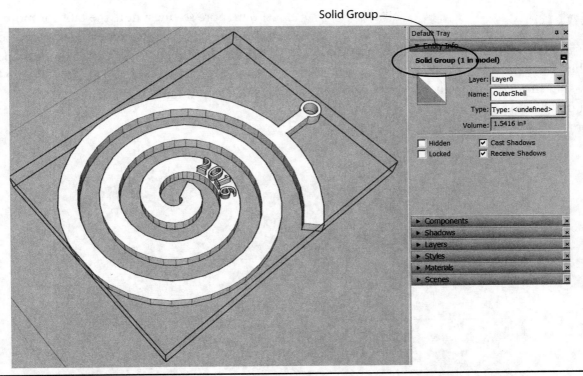

Figure 7-39 The spiral and numbers are one combined, solid group.

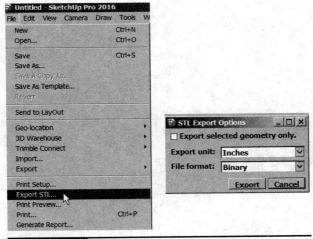

Figure 7-40 Export the ornament as an STL file.

Print It!

Figure 7-41 shows the build plate orientation on the MakerBot Mini and Simplify3D settings. Note that *Include Raft* is unchecked, which you can't do when using the MakerBot Desktop slicer for this particular printer. Figure 7-42 shows more settings; note the tabs for more setting options. The ornament was printed on a cold acrylic plate covered with painter's tape and wiped down with isopropyl alcohol. Figure 7-43 shows the printed ornament.

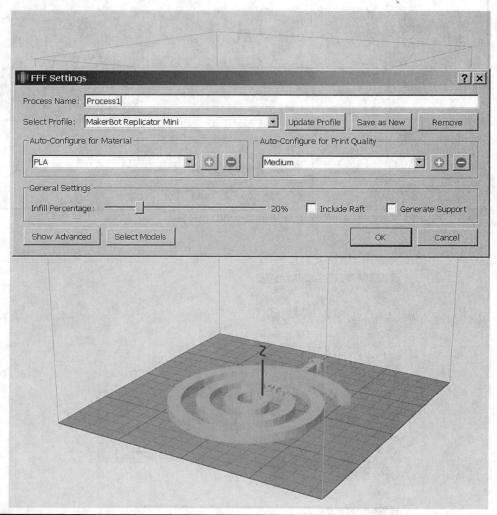

Figure 7-41   The file inside Simplify3D.

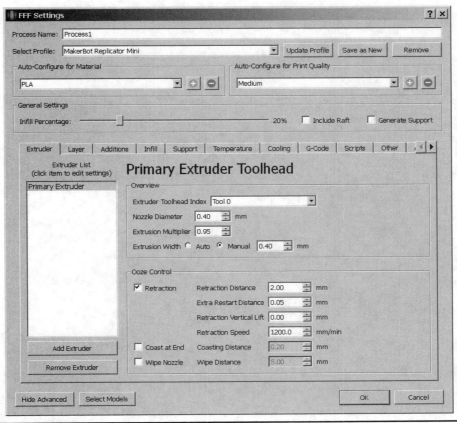

Figure 7-42 Simplify3D settings.

Figure 7-43 The printed ornament.

Personalized Football Key Fob

PROBLEM: A middle-schooler with limited digital modeling skills wants to personalize and print an STL file for a gift. We'll download a football key fob from Thingiverse (thingiverse.com/thing:609187), create text in Microsoft Paint, import and extrude the text in Simplify3D, and print the fob in PLA on a Gmax 1.5XT+.

Create Text

Launch Microsoft Paint, an app that's part of the Windows operating system. "Paintbrush" is the Mac equivalent. Type *mspaint* in the Start menu to find it (Figure 8-1).

Click on the Text tool (Figure 8-2). Choose a simple font with no serifs (they print best) and

Things You'll Need

Description	Source	Cost
Computer and Internet access	Your own or one at a makerspace	Variable
Microsoft Paint or Mac Paintbrush	Your computer's OS	Free
Simplify3D software	simplify3d.com/	$149
SketchUp Pro software	Sketchup.com	Free trial or $690
3D printer and slicing software	Your own, one at a public makerspace, or one at an online service bureau	Variable
Thumb drive (needed only for offsite printing)	Computer or electronics store	< $10
Spool of FLA filament	Amazon, Microcenter, or online vendor	Variable

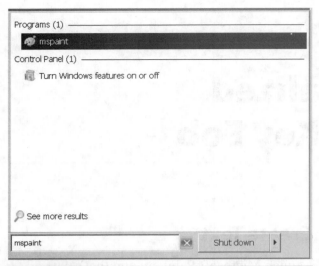

Figure 8-1 Type *mspaint* in the Start menu to find the Paint app.

a large font size. Then click on the workspace. Type your name inside the text box (Figure 8-3), and press ENTER. The text box will disappear, and the text will remain.

Crop out all the white space around the name (Figure 8-4). Click on the Crop tool, and then click Select and choose the Rectangular selection. Drag a rectangle around the name as close to it as possible, and then click the Crop tool again. The white space will be cropped out. Finally, click on the Paint menu, hover the mouse over Save As to get a popup submenu, and click on PNG (Figure 8-5).

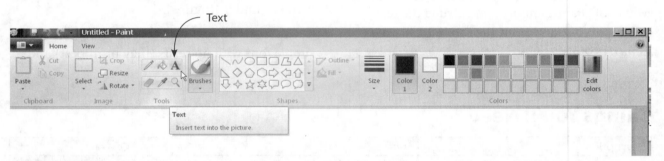

Figure 8-2 The Text tool.

Figure 8-3 Type your name.

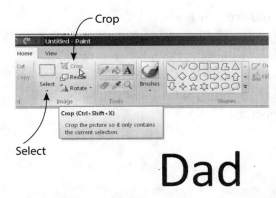

Crop

Select

1. Click the Crop tool.

2. Choose Rectangular selection.

3. Drag the selection window around the text and click the Crop tool again.

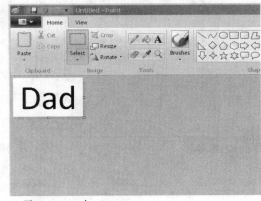

4. The cropped text.

Figure 8-4   Crop out the white space around the name.

Figure 8-5   Save the text as a PNG file.

Import the PNG File into Simplify3D

Launch Simplify3D (S3D). Click on Add-Ins and then on Convert Image to 3D (Figure 8-6). A navigation browser will appear; find the PNG file. On the browser box, click *Invert Depth Profile*, and then click Create. You'll be asked if you want to import the newly created model; click Yes (Figure 8-7). If it's too large for the build plate, Simplify3D will tell you so but will still import it.

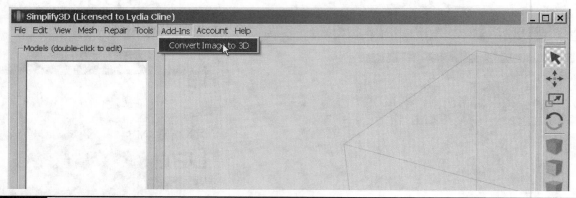

Figure 8-6   Click Add-Ins and Convert Image to 3D.

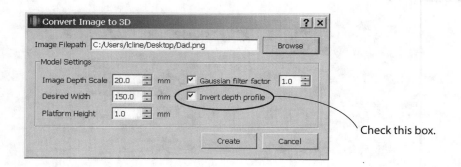

Check this box.

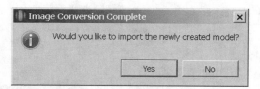

Figure 8-7   Navigate to the file, click *Invert Depth Profile*, and import the text model.

Size the PNG File Model

Double-click on the model to access a dialog box that has text fields to change the scale (Figure 8-8). Note the arrows in the build plate that show which dimension corresponds with which axis. Type the dimension you want in one field, and the others will adjust proportionately. If you only want to change one dimension, uncheck the Uniform Scaling box. Figure 8-8 shows the text file reduced along the *x*-axis from 77 to 49.99 mm.

Import the Football Key Fob STL File

Now import the football key fob STL file by clicking File/Import Models (Figure 8-9) or by simply dragging the STL file into the S3D workspace. Click on the Translate tool (Figure 8-10), select the text, and drag it onto the football (Figure 8-11). Position and push it a bit into the football. There's an Edit/Undo command at the top menu if you need to retrace your steps.

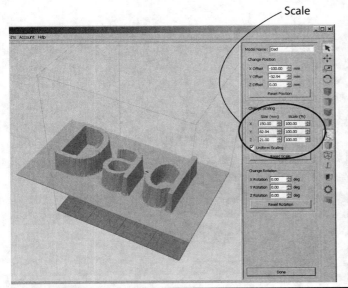
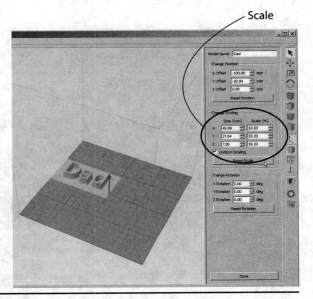

Figure 8-8 Change the scale of the text model as needed.

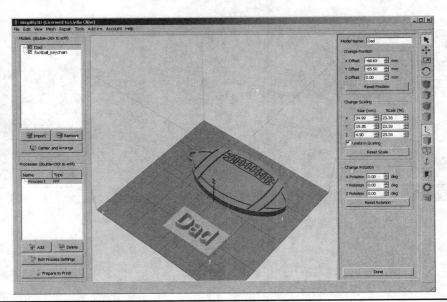

Figure 8-9 Import the football STL file.

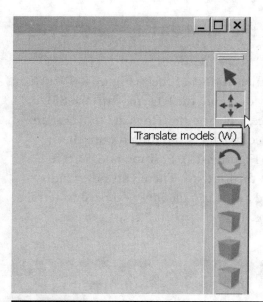

Figure 8-10 The Translate tool.

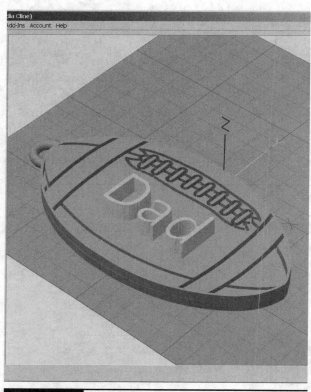

Figure 8-11 The text positioned on the football.

Print It!

Click on the Process 1 entry (Figure 8-12) to access the settings boxes. I chose a 15 percent

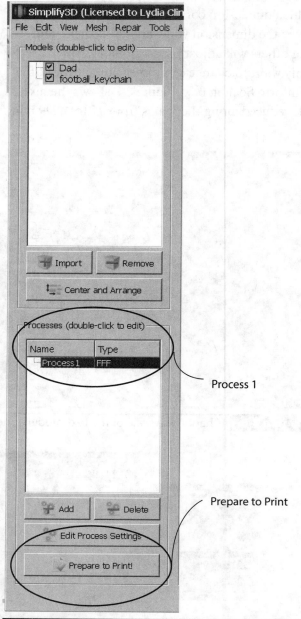

Figure 8-12 Click on the Process 1 entry to bring up the settings boxes.

infill and the settings shown in Figure 8-13. Then click on Save Toolpaths to Disk or Begin Printing over USB, whichever is appropriate, and a preview will appear (Figure 8-14).

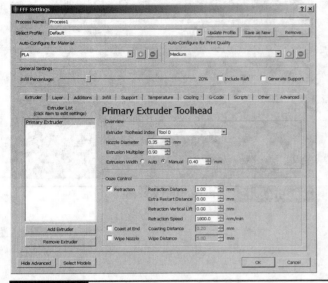

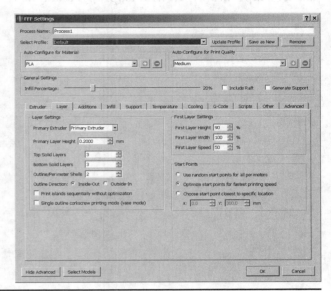

Figure 8-13 Some settings for this print.

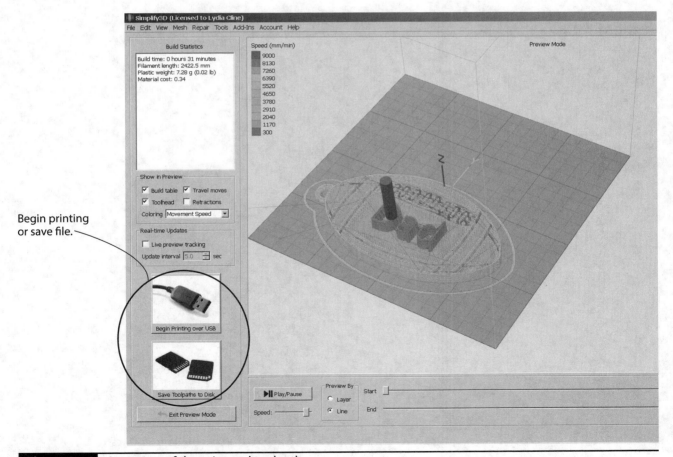

Begin printing or save file.

Figure 8-14 A preview of the print and toolpath.

Figure 8-15 shows the printed key fob. It was printed on a cold plate that was covered with painter's tape and wiped with isopropyl alcohol.

Figure 8-15 The printed key fob.

Embossed Poop Emoji Cookie Cutter

PROBLEM: A wise guy wants to bring some out-of-the-ordinary cookies to a party. In this project we'll use a downloaded PNG file, an online file converter, and Fusion 360 to make an embossed cutter. We'll slice it with Cura and print it with PLA on a Lulzbot Taz 6.

This cutter will be an outline plus details that emboss into cookie dough. A cutter can be modeled from a sketch you manually or digitally draw yourself or from one you find online and convert. We'll do the latter.

Find and Convert a Poop Emoji Sketch

Finding an existing sketch to model is often quicker than drawing one from scratch. Do a Google image search for "poop emoji" (Figure 9-1). A simple line drawing is best because elaborate details often convert poorly. PNG files are best because they preserve transparent areas in the drawing. Download one to your desktop.

Things You'll Need

Description	Source	Cost
Computer and Internet access	Your own or one at a makerspace	Variable
Autodesk account	autodesk.com/	Free
Autodesk Fusion 360 software	autodesk.com/products/fusion-360/overview	Free trial or subscription
3D printer and slicing software	Your own, one at a public makerspace, or one at an online service bureau	Variable
Thumb drive (needed only for offsite printing)	Computer or electronics store	< $10
Spool of PLA filament	Amazon, Microcenter, or online vendor	Variable

Figure 9-1 Results from a Google image search for "poop emoji."

After downloading an image, point your browser to image.online-convert.com. Click Convert image to the SVG format (Figure 9-2), navigate to the file, and upload it, keeping the default settings. Then click the Convert file button and download the SVG file (Figure 9-3).

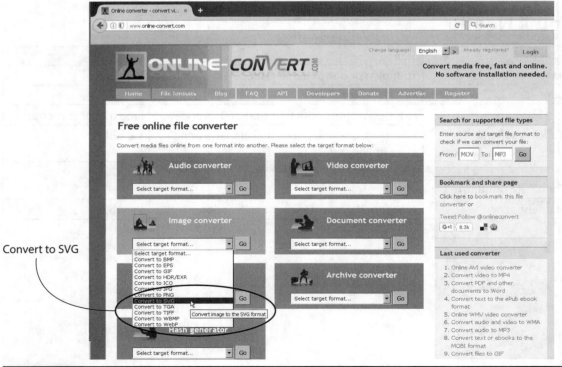

Figure 9-2 At image.online-convert.com, click Convert image to the SVG format.

Browse

Upload your image you want to convert to SVG:

Browse... No file selected.

Or enter URL of your image you want to convert to SVG:

(e.g. http://bit.ly/b2dIVA)

Or select a file from your cloud storage for a SVG conversion:

Choose from Dropbox   Choose from Google Drive

Optional settings

Change Size: ___ pixels x ___ pixels

Color: ● Colored ○ Gray ○ Monochrome ○ Negate
         ○ Year 1980 ○ Year 1900

Enhance: ☐ Equalize ☐ Normalize ☐ Enhance
          ☐ Sharpen ☐ Antialias ☐ Despeckle

DPI: ___

Convert file   (by clicking you confirm that you understand and agree to our terms)

Convert

Opening 2017-02-19_21-54-46.svg

You have chosen to open:

🗋 2017-02-19_21-54-46.svg

which is: scalable vector graphics file (3.1 KB)
from: http://www29.online-convert.com

What should Firefox do with this file?

○ Open with Inkscape (default)
● Save File

☐ Do this automatically for files like this from now on.

OK   Cancel

Figure 9-3 Convert and download the file.

Import the SVG File into Fusion 360

Launch Fusion 360. On the menu at the top of the screen, click Insert/Insert SVG (Figure 9-4). A dialog box and three planes will appear (Figure 9-5). Click on the file folder, and navigate to the SVG file. Then click on the horizontal plane. The file will enter (Figure 9-6). You can drag the arrows to move it or change its size by typing new dimensions in the text field; I typed *127* in the *y* field to make this big enough for a cookie-cake.

INSERT ▼   MAKE ▼   ADD-INS ▼   SELEC

🔲 Decal
🖼 Attached Canvas
📥 Insert Mesh
📥 Insert SVG
📥 Insert DXF
📥 Insert McMaster-Carr Component
📥 Insert a manufacturer part

Figure 9-4 Click Insert/Insert SVG.

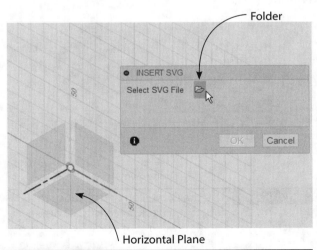

Folder

INSERT SVG

Select SVG File 🗁

OK   Cancel

Horizontal Plane

Figure 9-5 Click on the folder to navigate to the SVG file, and then click on the horizontal plane.

Figure 9-6 The imported SVG file.

Return the workspace to a 3D display by either clicking on the Orbit tool or hovering the mouse over the View Cube to make the house appear and then clicking on the house (Figure 9-7).

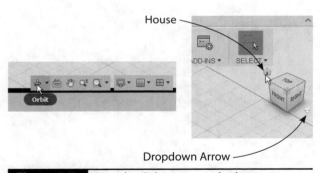

Figure 9-7 Use the Orbit icon or the house to return the workspace to a 3D display.

Examine the File for Problems

Tip: You never know how well a picture from the Web will convert to another format, especially when it contains colors and details. You also don't know how well it will import into a particular software program. Examine the sketch to see if modifications are needed before extruding it, such as adding lines to separate areas that will currently extrude together (if you don't want them to extrude together), trimming lines that will currently prevent areas from extruding together (if you want them to extrude together), or extending lines to close gaps. Tools to do that are in the Sketch menu.

Offset the Perimeter Sketch

We need to give thickness to the perimeter line, so click on Sketch/Offset (Figure 9-8). Then drag the arrow or type the thickness you want. I typed *4 mm* (Figure 9-9).

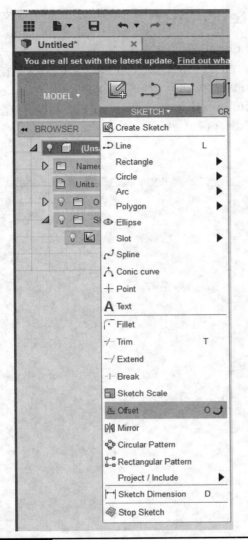

Figure 9-8 Click on Sketch/Offset.

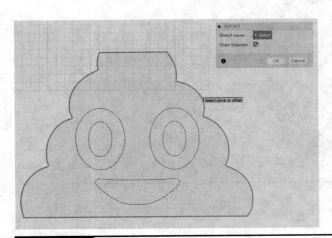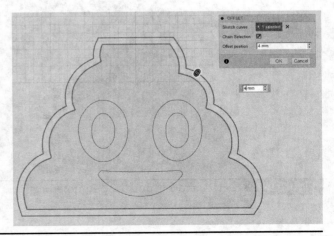

Figure 9-9 Offset the perimeter.

Extrude the Perimeter Sketch

Return to a 3D view by clicking on the House icon or Orbit. Then right-click on the perimeter, choose Press Pull from the options tree, and extrude the perimeter of the sketch-up (Figure 9-10). The interior sketches will disappear; in the Browser box, click the dropdown arrow next to Sketches, and find the sketch(es) that were turned off. They'll have a gray bulb in front of them. Click on the bulb to turn it yellow (Figure 9-11). If multiple sketches are turned off, select them all by holding the SHIFT key down.

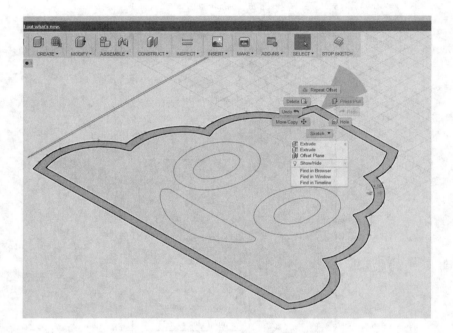

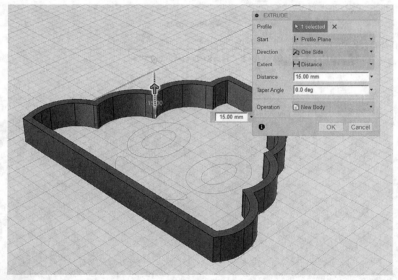

Figure 9-10 Select and extrude the perimeter up.

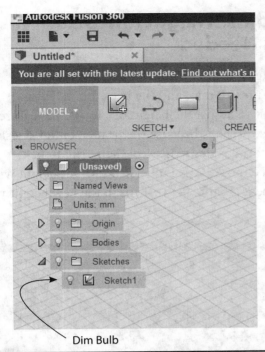

Dim Bulb

Figure 9-11 Turn the interior sketches back on by clicking on the gray bulb.

Extrude the Interior Sketches

Click on all the interior sketches (hold the SHIFT key down to make multiple selections). Right-click on one selected sketch, click on Press Pull, and extrude the sketch up a lower height than the perimeter (Figure 9-12). Those sketches will create the embossing.

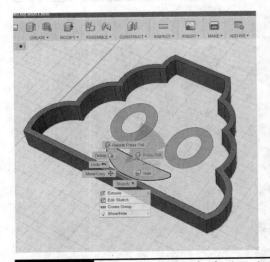
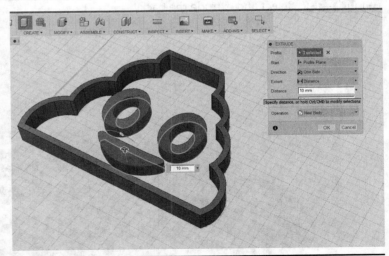

Figure 9-12 Select and extrude the interior sketches up.

Extrude the Back

We need to extrude the back of the sketch to hold all the parts together. In the Browser box, click on the same sketch listing again. The back will highlight. Right-click anywhere on it, click on Press Pull, and extrude it up (Figure 9-13).

Adjust Height and Size as Needed

I raised the perimeter 15 mm, the interior features 10 mm, and the back 5 mm. If you want to adjust the height of any of the features later, perhaps because the printed cutter doesn't perform as desired, select them, right-click, and choose Press Pull. Their current height will appear (Figure 9-14). Type in the new height you

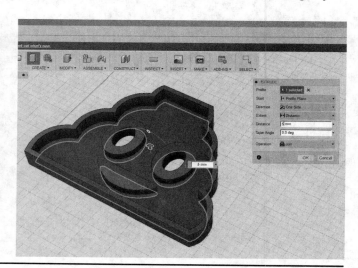

Figure 9-13   Select and extrude the back.

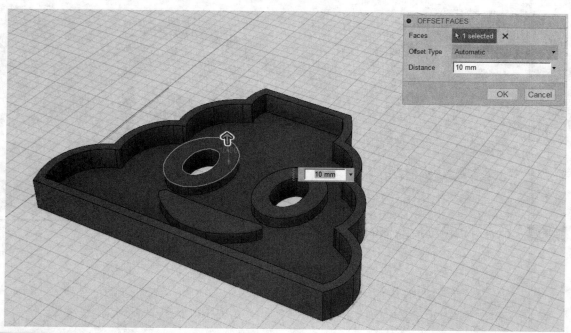

Figure 9-14   Readjust features by selecting and typing a new height.

want—the total height, not an addition to the current height. You can also change the cutter's entire size. To find out its current size, click on Inspect/Measure, and then click on the first

icon in the Measure dialog box (Figure 9-15). Click on opposite ends of the cutter to find the distance between them (Figure 9-16).

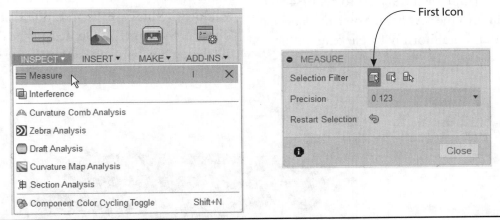

Figure 9-15   Click on Inspect/Measure and then on the first icon in the dialog box.

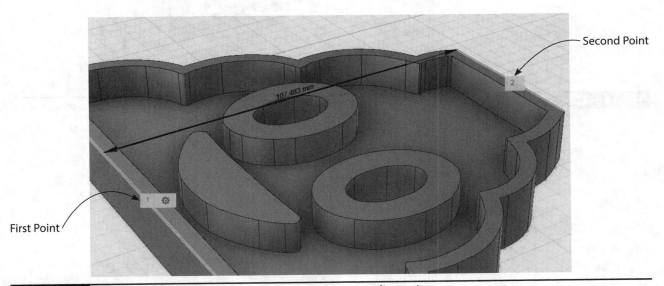

Figure 9-16   Click on opposite ends to find the cutter's current dimension.

To change its current size, click Modify/ Scale and either drag the arrow to eyeball a new size or click a scale factor into the dialog box (Figure 9-17). To scale the cutter to a specific size, divide the size you want by the size you have. The cutter is currently 107 mm between the points measured. If you want that distance to be, for example, 90 mm, divide 90/107 to get 0.84, and type that into the text field. You can also scale the cutter nonuniformly by clicking on the dropdown arrow next to Uniform in the Scale dialog box.

Export the Cookie Cutter as an STL File

Right-click on the file's name in the Browser window (in this case, Poop Emoji Cutter v1). Click on Save as STL file, accept the defaults in the dialog box, and click OK (Figure 9-18).

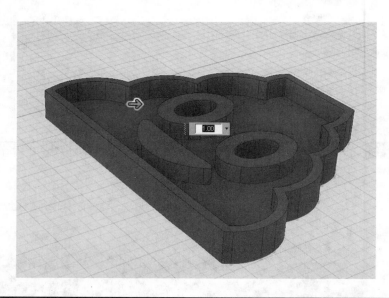

Figure 9-17   Click Modify/Scale, and drag the arrow or type a number in the Scale Factor text field to scale the cutter.

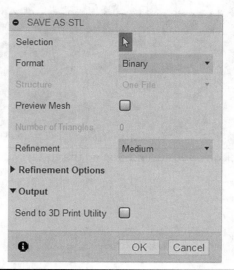

Figure 9-18   Right-click on the file's name and click on *Save as STL file*.

Print It!

I imported the file into Cura and laid it flat, face up, on the build plate (Figure 9-19). I gave it a 15 percent infill. The cutter was printed on a hot borosilicate glass build plate covered with PEI. Figure 9-20 shows the final print.

Figure 9-19 Settings and orientation in Cura.

Figure 9-20 The printed cookie cutter.

Lens Cap Holder

PROBLEM: A photographer needs an attachment to hold the camera's lens cap while taking photos (Figure 10-1). We'll use Fusion 360 to design a cap holder that fits onto the camera's strap. Then we'll slice it with Cura and print it with PLA on a Lulzbot Taz 6.

Figure 10-2 shows the lens cap. It attaches to the camera lens with spring-loaded clips and will need to attach to a holder the same way. When designing an item to fit with other items, use a digital caliper to get exact dimensions. The cap's diameter is 43.16 mm, and the strap on which it will be placed is 22 mm wide.

Things You'll Need

Description	Source	Cost
Computer and Internet access	Your own or one at a makerspace	Variable
Autodesk account	autodesk.com/	Free
Autodesk Fusion 360	autodesk.com/products/fusion-360/overview	Free trial or subscription
3D printer and slicing software	Your own, one at a public makerspace, or one at an online service bureau	Variable
Thumb drive (needed only for offsite printing)	Computer or electronics store	< $10
Spool of PLA filament	Amazon, Microcenter, or online vendor	Variable

Figure 10-1 Camera and lens cap.

Figure 10-2 Use a digital caliper to get exact measurements.

Sketch the Holder

Launch Fusion 360. On the menu at the top of the screen, click on Sketch/Circle/Center Diameter Circle (Figure 10-3). Of course, if the circle appears on the menu, just click on it. Then click on the horizontal plane, click on the origin for the circle's center point, and type *43.13* (the units are mm) for the diameter (Figure 10-4).

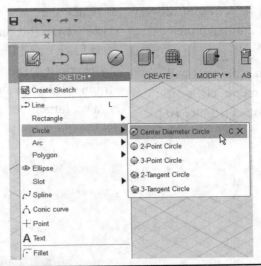

Figure 10-3 On the menu at the top of the screen, click on Sketch/Circle/Center Diameter Circle.

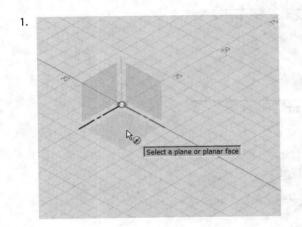

1.

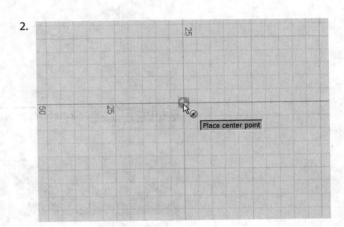

2.

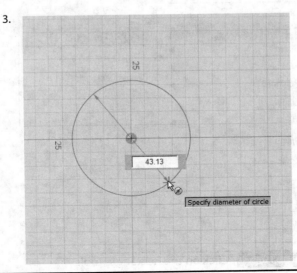

3.

Figure 10-4 Sketch a circle.

Sketch a second circle. Click its center at the same origin point, and make it 2 mm smaller than the first one (Figure 10-5).

Now we need to sketch the handle that will attach to the strap. On the top menu, click on Sketch/Rectangle/2-Point Rectangle (Figure 10-6).

Click the first point in the upper-left corner, type *20 mm* (hit the TAB key to toggle between text fields), and then click the second point in the lower-right corner. The handle width should be thinner than the strap width to stay in place. Then draw a larger rectangle around it (Figure 10-7).

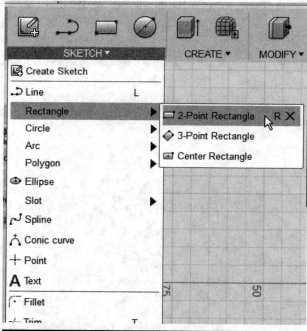

Figure 10-5   Sketch a second, smaller circle.

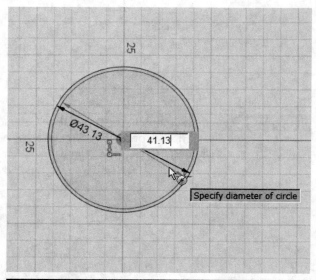

Figure 10-6   Click on the Rectangle tool.

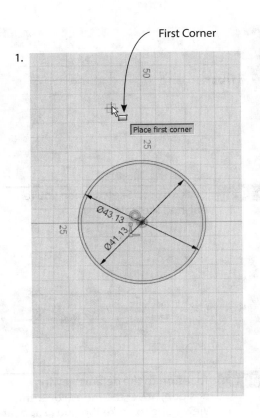

1. First Corner

Place first corner

Ø43.13

Ø41.13

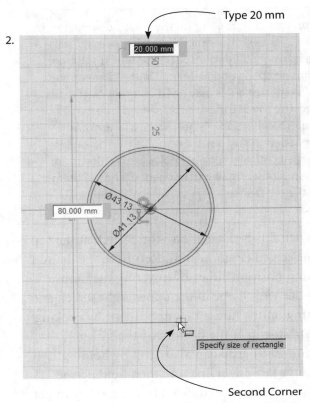

2. Type 20 mm

20.000 mm

80.000 mm

Ø43.13

Ø41.13

Specify size of rectangle

Second Corner

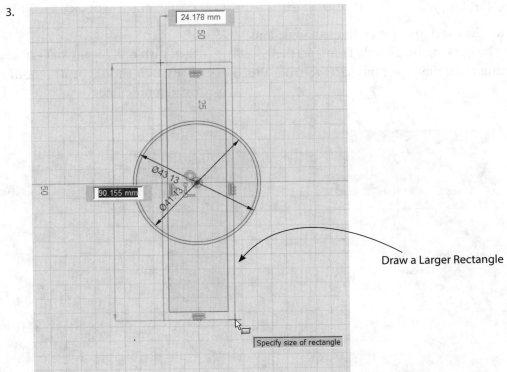

3.

24.178 mm

Ø43.13

Ø41.13

90.155 mm

Draw a Larger Rectangle

Specify size of rectangle

Figure 10-7   Sketch a 20-mm-wide strap handle.

Tip: A good start point for the circle's diameter is 2 to 3 mm, but the exact dimension needed will be influenced by your specific printer, resolution, printing speed and filament. On the same printer, different size holes may need different clearances because the STL mesh circle is not a true curve but rather a collection of lines. Some trial and error is needed to find a size that holds the cap on the strap. If critically accurate holes are needed, print them slightly smaller than needed, and file them to the exact size.

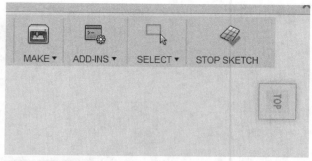

Figure 10-8 Click on Stop Sketch when you're finished sketching.

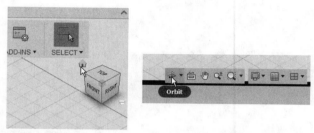

Figure 10-9 To return to the default viewing position, hover the mouse over the View Cube to make the house appear or click the Orbit icon.

When finished, click on Stop Sketch at the top of the screen (Figure 10-8). Return to the default viewing position either by hovering the mouse over the View Cube in the upper-right screen and clicking on the house that appears or by clicking the Orbit icon at the bottom of the screen (Figure 10-9). You can also orbit by holding the SHIFT key down first and then holding the mouse's scroll wheel down. Hit the ESC key to get out of Orbit mode.

Trim the overlap between the rectangle and circles by clicking on Sketch/Trim and then hovering the mouse over the parts to trim. When they turn red, hit the DELETE key (Figure 10-10). The result should be a smooth sketch with no overlapping lines.

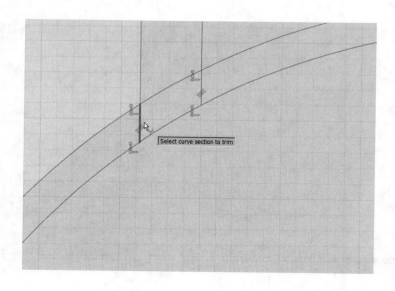

Figure 10-10 Trim overlapping lines.

Extrude the Sketch

Hold the SHIFT key down, and click on the circle and handle outlines to select them. Then right-click to bring up the options tree and click Press

Pull. Type *4 mm* in the box that appears, or drag the arrow up until the text box shows 4 mm (Figure 10-11).

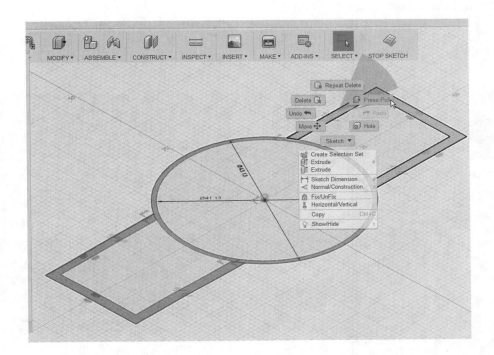

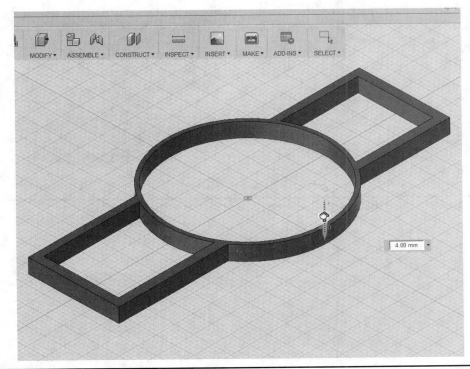

Figure 10-11 Select and extrude the sketch up 4 mm.

Sketch and Extrude a Second Circle

We need to sketch another circle to serve as the lens holder's back. Click on Sketch/Circle/ 2-Point Circle, and then click on the horizontal plane. Then click on the origin for the center and 18 mm for the diameter (Figure 10-12). Return to the default view, select the new circle sketch, right-click, click on Press Pull, and extrude it up 2 mm (Figure 10-13).

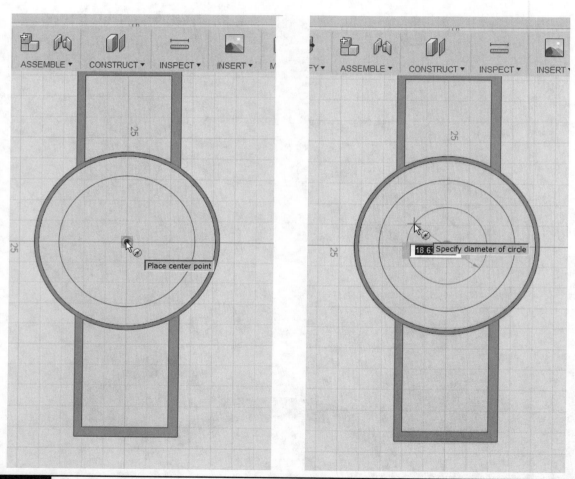

Figure 10-12 Sketch a circle that shares the same center point as the existing circle.

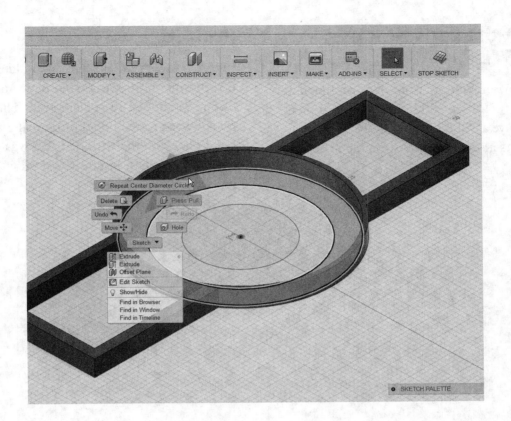

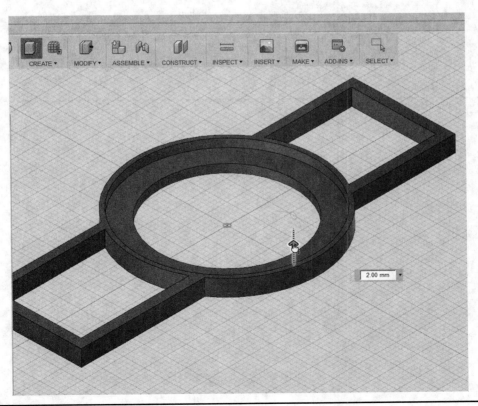

Figure 10-13 Extrude the circle up 2 mm.

Export the Lens Holder as an STL File

In the Browser window, right-click on the name of the file (in this case "Untitled 2"), and then click on Save as STL file (Figure 10-14). Accept the defaults in the dialog box that appears, and click OK.

Print It!

Figure 10-15 shows the lens cap holder's orientation on the Cura build plate and some settings. I chose a 15 percent infill. The lens cap holder was printed on a hot borosilicate glass build plate covered with PEI that was wiped with isopropyl alcohol. Figure 10-16 shows the print on the build plate and in use.

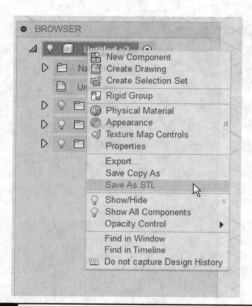

Figure 10-14   Save the file as an STL file.

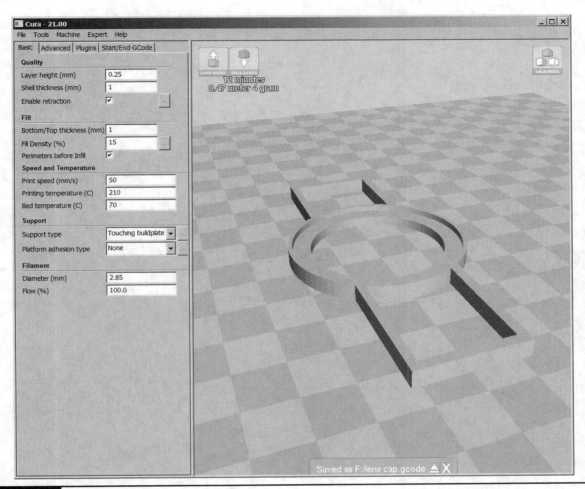

Figure 10-15   The file in Cura.

Figure 10-16   The printed lens cap holder.

Print Two Colors with a Dual Extruder

PROBLEM: A teacher wants to demonstrate making a two-color print with a dual-extruder printer. In this project we'll prepare a model for two-color printing and use Simplify3D's (S3D) Dual Extrusion Wizard to slice it. Then we'll print it in PLA on a Gmax 1.5XT+ printer.

The S3D slicer lets you make a dual-color print with a dual extruder two ways. One is to stop a print at the layer where a color change starts and have the second extruder take over.

You only need one STL file for this. The other is to use the Dual Extrusion Wizard to tell the extruders to stop and start as needed. Two STL files are needed for this. Both methods require setting up a *process* for each extruder, which is a set of slicing instructions.

We're going to use SD3's Dual Extrusion Wizard. We'll import a model into Fusion 360 and cut it into two files, and export each as an STL file.

Things You'll Need

Description	Source	Cost
Computer and Internet access	Your own or one at a makerspace	Variable
Autodesk account	autodesk.com/	Free
Fusion 360 software	autodesk.com/products/fusion-360/overview	Free or subscription cost
Simplify3D software	Simplify3d.com	$149
3D printer and slicing software	Your own, one at a public makerspace, or one at an online service bureau	Variable
Thumb drive (needed only for offsite printing)	Computer or electronics store	< $10
Spool of PLA filament	Amazon, Microcenter, or online vendor	Variable
Black 1/8"-wide Chartpak tape (optional)	Amazon, hobby shop	< $5

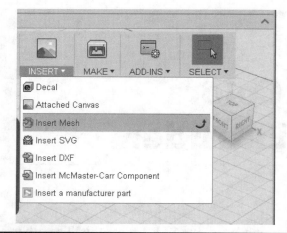

Figure 11-1 Import the STL file at Insert/Insert Mesh.

Import and Convert the STL File

Click on Insert/Insert Mesh to import the STL file into Fusion 360 (Figure 11-1). If the Sculpt mode appears on import, click on the dropdown arrow to change the mode to Model (Figure 11-2). Next, convert the file to a solid so that we can work on it. In the browser window, right-click on the file name (in this case "Bowl Model v1"), and click on Do Not Capture Design History (Figure 11-3). A warning dialog box will appear; click Continue. Then right-click on

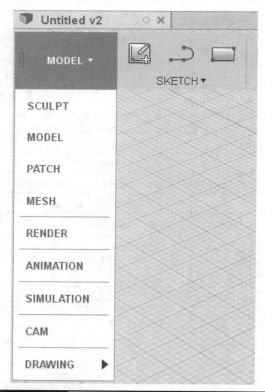

Figure 11-2 If Sculpt mode appears on import, change it to Model.

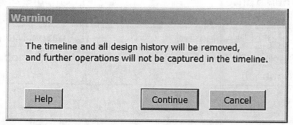

Figure 11-3 Click on Do Not Capture Design History.

the model to select it, and click Mesh to BRep (Figure 11-4). The STL file will convert to a solid. Know that this conversion is only possible if the STL mesh has no holes or other problems; if it does, an error message will appear. You might also get a warning if the STL file is large, but click OK anyhow because Fusion may still be able to convert it.

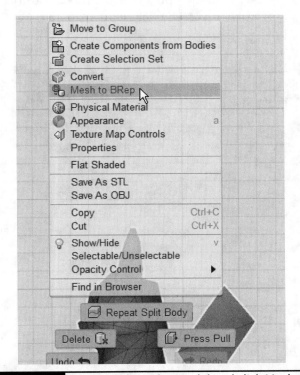

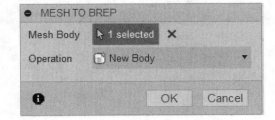

Figure 11-4   Right-click on the model and click Mesh to BRep.

Split the Ears from the Body

The ears and the body will be different colors, so they must be separate STL files. We'll draw a splitting line and then use that line to split the model.

Make a Box

Click on Create/Box, and drag a box into the workspace. Right-click two corners to size the box, and use the arrows to pull it up the same height or higher as the top of the model (Figure 11-5). It doesn't matter if the bottom of the box is level with the bottom of the model or not.

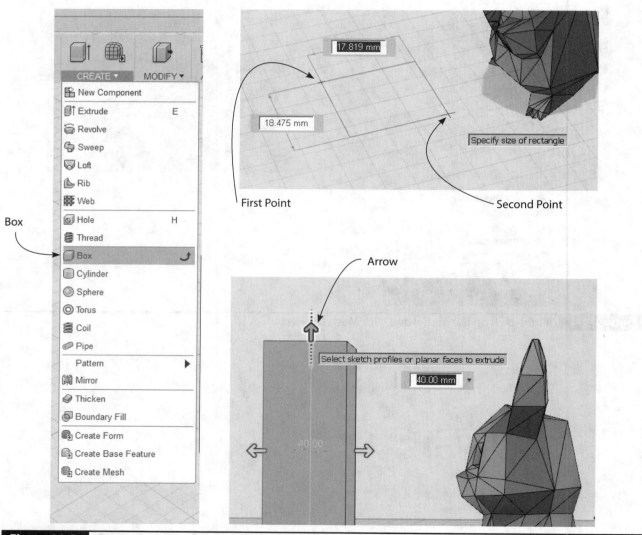

Figure 11-5   Bring a box into the workspace, and pull it up level with the top of the model.

Sketch a Line

Hover the mouse over the View Cube to make a dropdown arrow appear, and then click on Ortho mode (Figure 11-6). This will make sketching easier. Then click on Sketch/Line and draw a line on the box at the location you want the second color to start (Figure 11-7). If you have trouble sketching the line perfectly horizontal, type *0* in the degree text field. Hit the TAB key to switch between the length and text fields.

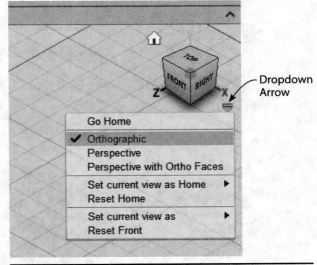

Figure 11-6   Click on Ortho mode.

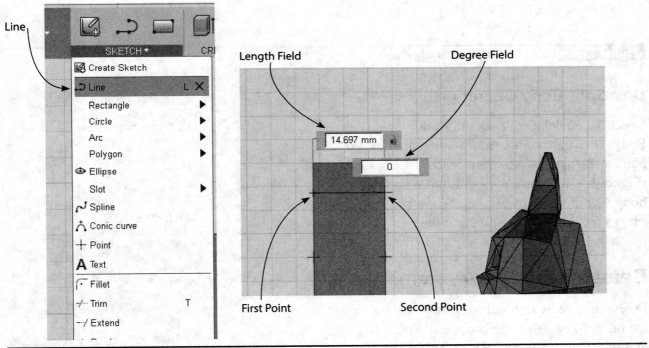

Figure 11-7   Sketch a line on the box.

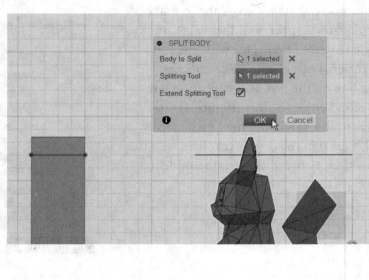

Figure 11-8   Split the ears from the body.

Do a Split-Body Operation

Select the model, then click on Modify/Split
Body. Then click on the line you just drew. A red
plane will appear at the split location. Click OK
(Figure 11-8), and the ears will split from the
body. Confirm this by clicking on the body and
ears to see if they highlight separately.

Delete Unnecessary Items

Delete the line by right-clicking on it and hitting
the DELETE key. Select the box, right-click,
and then click on Show/Hide (Figure 11-9).
Alternatively, find its listing in the Browser box,
and click on the yellow light bulb in front of it to
hide it (Figure 11-10).

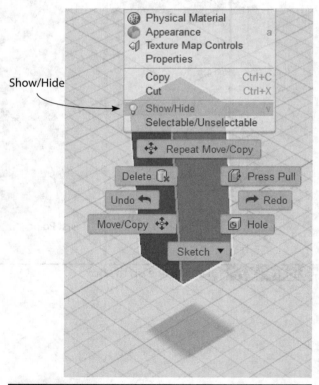

Figure 11-9   Hide the box by right-clicking on it
and clicking Show/Hide.

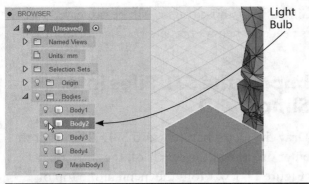

Light Bulb

Figure 11-10 **Figure 11-10**   The box can also be hidden by right-clicking on its browser listing and clicking the yellow light bulb to make it dim.

Export Two STL Files

Export the ears and body as separate STL files. First, right-click on the body's browser listing, and click Show/Hide (Figure 11-11). Only the ears will remain visible (you'll see a ghost outline of all the hidden items). Click Save as STL file to export the ears as an STL file, and accept the defaults in the dialog box that appears. Call the file "Ears."

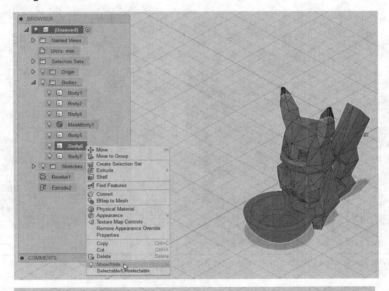

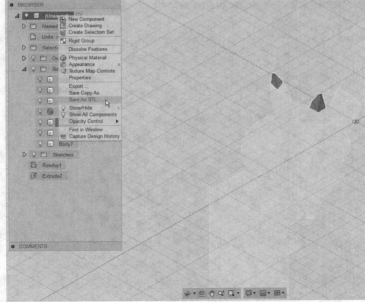

Figure 11-11   Hide the body and export the ears as an STL file.

Now hide the ears (note that there are two listings for them in the Browser window). Again, the hidden items show up as ghost outlines. Save the body as an STL FILE, and call it "Body" (Figure 11-12).

It is important to keep the body and ears in their current positions when exporting them as STL files because their origins will be exported with them. When we import them into Simplify3D, those origins will be preserved, which is needed when we prepare the files for printing.

Import the STL Files into Simplify3D

Drag the files separately into S3D. They'll both enter on, and at the center of, the work plane (Figure 11-13). From the menu at the top of

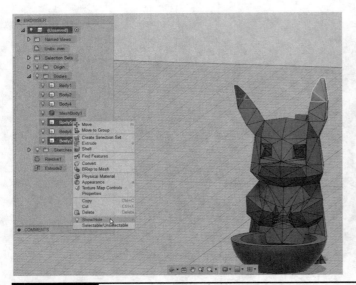
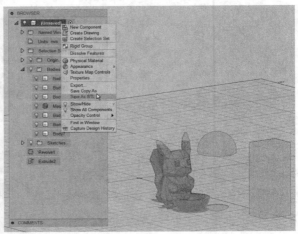

Figure 11-12   Hide the ears, and export the body as an STL file.

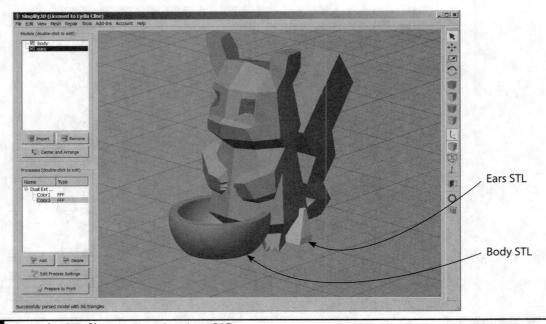

Ears STL

Body STL

Figure 11-13   Drag the STL files one at a time into S3D.

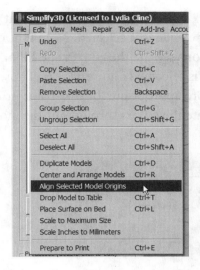
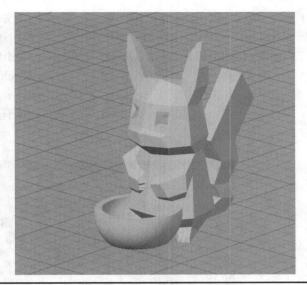

Figure 11-14 Click Edit/Align Selected Model Origins to snap the ears on top of the body.

the screen, click on Edit/Align Selected Model Origins. This will snap the ears to their origin, which is on top of the body (Figure 11-14).

Set Up Processes

Click the Add button in the Processes box twice to set up two processes (slicer settings), one for each STL file (Figure 11-15). Click on each process to choose all its settings. The settings should be the same for each process, the difference being that one process will be for

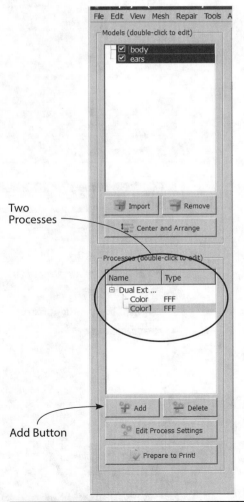

Figure 11-15 Make a process (slicer settings) for each STL.

the left extruder and one will be for the right extruder (Figure 11-16). Note that in the Auto-Configure Extruders field, *Both Extruders* is chosen.

Click on each tab to make sure that the proper extruder is selected. Note that under the Additions tab, neither the Prime Pillar nor Ooze Shield is selected (Figure 11-17). Those features are more appropriate for two-color prints that alternate the colors throughout the model.

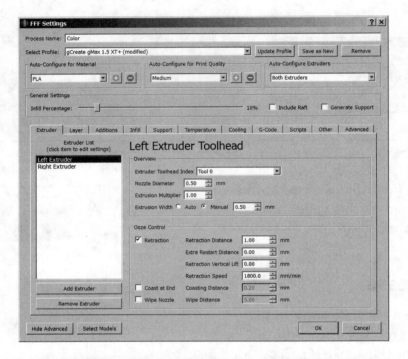

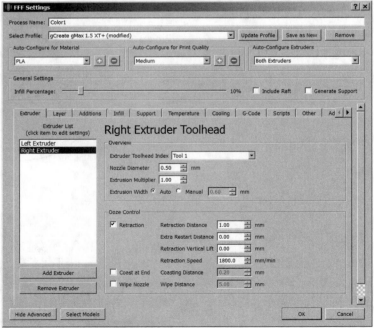

Figure 11-16 A process for the left extruder and a process for the right extruder.

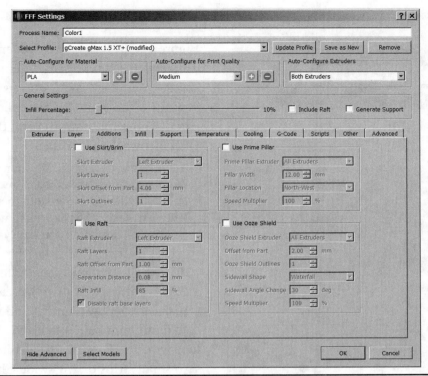

Figure 11-17 On the Additions tab, the Prime Pillar and Ooze Tower options are unchecked.

The Dual Extrusion Wizard

Now click Tools/Dual Extrusion Wizard (Figure 11-18). A box will appear in which to assign an extruder to an STL file. Finally, you'll be asked to select the processes used for the print; hold the SHIFT key down to select both entries, and click the Continuous Printing button (Figure 11-19).

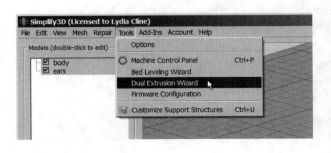

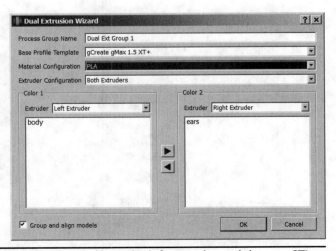

Figure 11-18 Click the Dual Extrusion Wizard and assign the body STL file to the left extruder and the ears STL file to the right extruder.

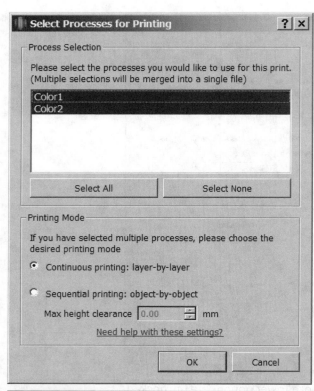

Figure 11-19   Select both processes.

Print It!

Figure 11-20 shows the extruder paths (the different colors represent different extruder speeds), and the *Build Statistics* box shows the amount of time and filament the model needs. Choose how you want to print the model—over USB or by SD card (disk).

> *Tip:* Hold a level to the dual extruders after adjusting them to ensure that they are indeed level with each other. Extruders at different heights will not print successfully (Figure 11-21).

Figure 11-22 shows the printed model. It has a 10 percent infill that can be seen through the translucent yellow filament and was printed on a cold aluminum build plate, over which painter's tape wiped with isopropyl alcohol was placed. Two small strips of 1/8"-wide Chartpak tape were added for the eyes.

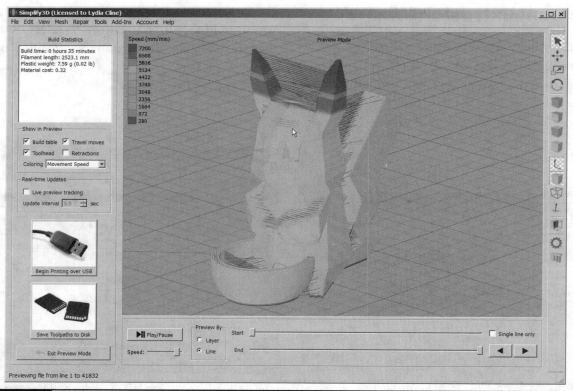

Figure 11-20   The extruder tool paths.

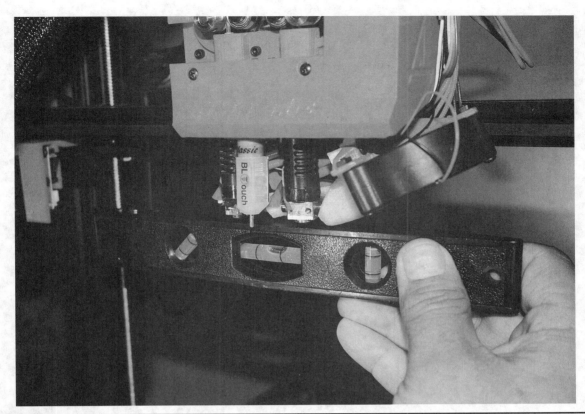

Figure 11-21 Both extruders must be perfectly aligned with each other.

Figure 11-22 The printed model.

Lithophane Night-Light

PROBLEM: A backpacker wants to turn a favorite photo into a night-light cover. In this project we'll use an online converter to turn a photograph into a lithophane, slice it with MakerBot Desktop, and print it with PLA on a MakerBot Mini+ printer.

Find an Image

A *lithophane* is an image etched in thin, translucent porcelain through which light can shine. To 3D-print a lithophane, an appropriate image is needed. It can be a photograph, sketch, or cartoon, but a simple, high-contrast one

Things You'll Need

Description	Source	Cost
Computer and Internet access	Your own or one at a makerspace	Variable
Autodesk account	autodesk.com/	Free
Autodesk Meshmixer software	Meshmixer.com	Free
3D printer and slicing software	Your own, one at a public makerspace, or one at an online service bureau	Variable
Thumb drive (needed only for offsite printing)	Computer or electronics store	< $10
Spool of PLA filament	Amazon, Microcenter, or online vendor	Variable
STL file for connector piece	thingiverse.com/thing:2024849	Free
GE incandescent night-light, #52194	Amazon.com	< $5
Loctite gel glue	Amazon or discount store	< $5

Figure 12-1   A high-contrast image converts best.

(Figure 12-1) with a uniform background works best because fine details may get lost in the conversion or printing. If your image doesn't convert well, import it into Photoshop and adjust the brightness/contrast. If it's a color image, turn it into a gray scale image (Figure 12-2). If you want a built-in frame, leave a border around it when you crop it, or make a border with Photoshop's crop tool.

Frame added with crop tool   Brightness/Contrast Adjustment

Crop tool

Figure 12-2   In Photoshop you can add a frame with the crop tool, convert a color image to a gray scale image, and adjust the brightness/contrast.

Convert the Image to a Lithophane

Point your browser to 3dp.rocks.com. Click on Tools, and on the Tools page, click on Image to

Lithophane Converter (Figure 12-3). The screen in Figure 12-4 will appear. A flat plane surface is the default, but you can click on any of the options.

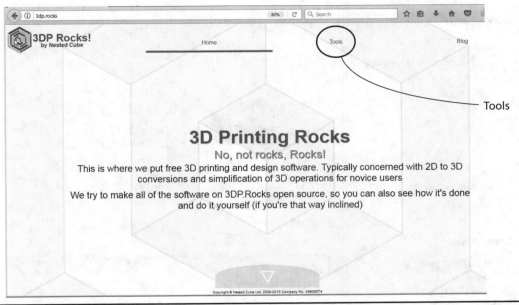

Tools

Figure 12-3   At 3dp.rocks.com, click on Image to Lithophane Converter.

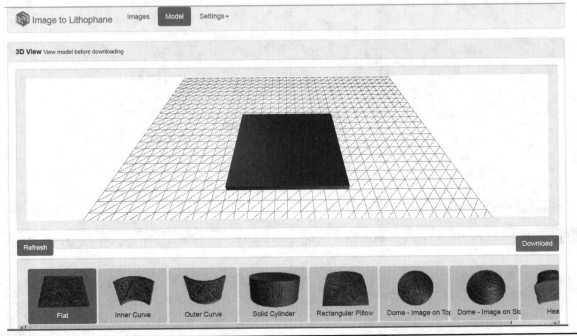

Figure 12-4   The flat plane is the surface default. Other options are shown at the bottom.

Next, click on the Images button at the top of the page. A navigation browser will appear.

Import the file, and click ENTER. The image will appear in the converter (Figure 12-5).

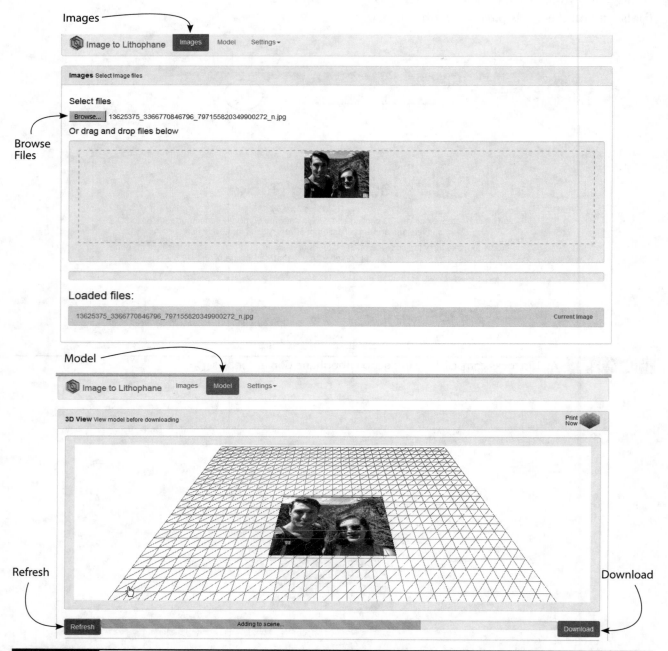

Figure 12-5   Click on the Images button for a navigation browser (*top*), and then import the file (*bottom*).

Before going further, click on the Settings button at the top of the screen (Figure 12-6). There are options for Model, Image, and Download. Click on the Model settings (Figure 12-7). The first slide bar is set to Negative as a default, which etches the picture into the background. Click on Positive, which raises the lines above the background. A 5-mm or thinner lithophane lets the most light through. Then click on the Model button to return to the image, and click the Refresh button at the bottom-left corner to see the changes (Figure 12-8).

Figure 12-6 Click on the Settings button for model, image, and download options.

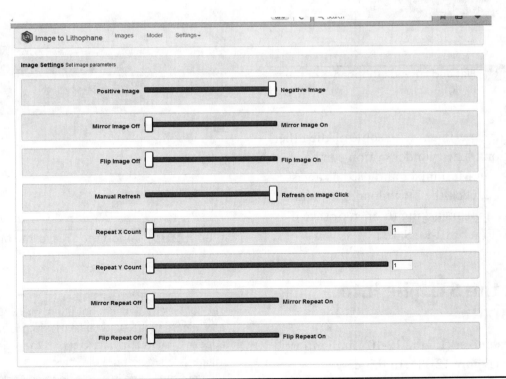

Figure 12-7 The top slider sets the lithophane positive or negative.

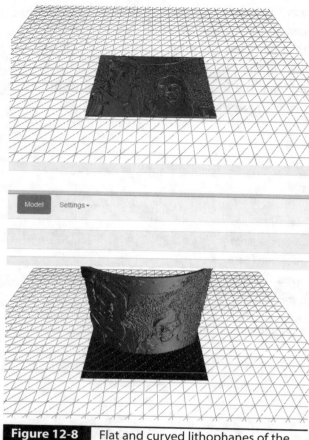

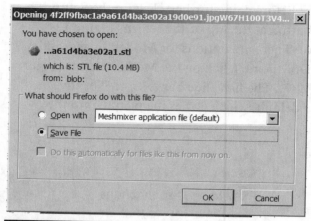

Figure 12-9 Download and save the STL file.

Figure 12-8 Flat and curved lithophanes of the image.

You might experiment with some of the other surface options. The cylinder option can be hollowed out in Meshmixer and used as a vase. When finished, click the Download button and a dialog box appears asking to open or save it as an STL file (Figure 12-9). Save it.

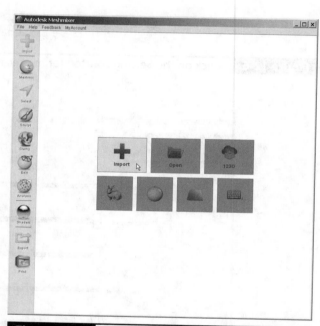

Figure 12-10 Import the STL file into Meshmixer.

Import the STL File into Meshmixer

Launch Meshmixer (Figure 12-10), and import the STL file (Figure 12-11). In this case, it imported in the orientation in which we want to print it (vertical), so there's no need to reorient it.

> *Tip:* For files that need reorienting after import, press and release the T key to make the Transform tool appear. Drag its arrows and curves to reorient the image. Drag the mouse directly over the radial marks to orient the image in whole degrees. Click Accept when finished.

Figure 12-11 The imported lithophane file.

Change the Size (Optional)

Click on the Analysis icon. This accesses a Tools menu; click on Units. Dimensions showing the model's size will appear (Figure 12-12). If you want to change the size, type a new dimension in one of the text fields and then click on the workspace. The other dimensions will adjust proportionately. Then click Done.

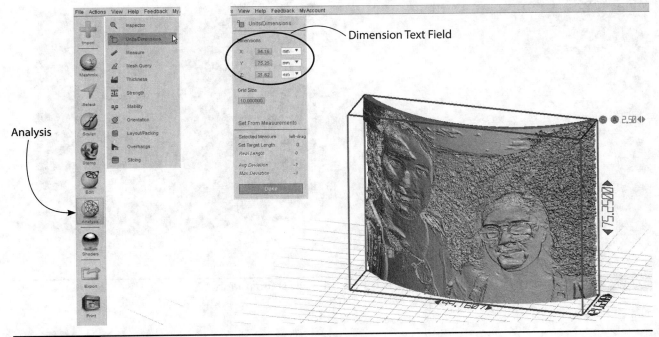

Figure 12-12 At Analysis/Units, verify or change the file's size.

While in the Analysis mode, click on Stability. The green ball that appears means that the print will stand upright without teetering (Figure 12-13). This is nice if you plan to just display this on a table. Finally, click the Export icon on the vertical menu bar and export the file as an STL file.

Print It!

For best results, print the lithophane vertically. It may do better on a Core XY printer than on a RepRap design because the momentum of the build plate moving back and forth on the RepRap is often too much for the small surface area to hold. If you must use a RepRap, orienting the model parallel to the bed movement helps a little. Don't set the speed faster than 50 mm/s.

Figure 12-14 shows the model's orientation on the build plate, and Figure 12-15 shows the settings. Note that supports are turned off. I printed it with white filament and a 5 percent infill. Hold a flashlight behind the print while it's under construction to see how it looks. If you're not getting the result you want, end the print and change the settings. The print was made on a cold acrylic build plate with painter's tape wiped with isopropyl alcohol.

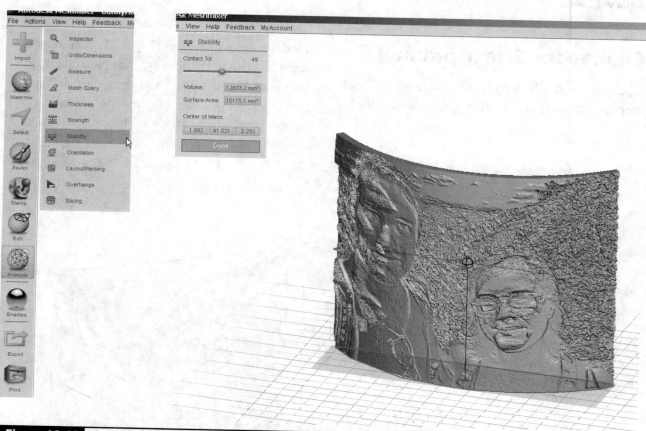

Figure 12-13   At Analysis/Stability, check to see if the print will stand upright.

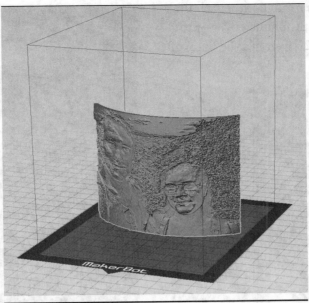

Figure 12-14 The model's orientation on the build plate.

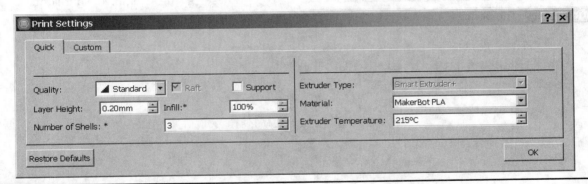

Figure 12-15 The settings.

Assemble It

I bought a GE incandescent night-light (#52194). The STL connector piece file at thingiverse.com/thing:2024849 (Figure 12-16) is designed to attach to it. Figure 12-17 shows all the pieces you need, including the printed connector piece. Glue the lithophane to the

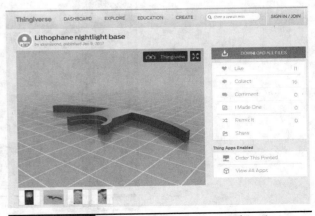

Figure 12-16 The connector piece at Thingiverse.

connector, remove the cover that comes with the night-light, and snap the lithophane cover onto it (Figure 12-18). Figure 12-19 shows the back of the new night-light.

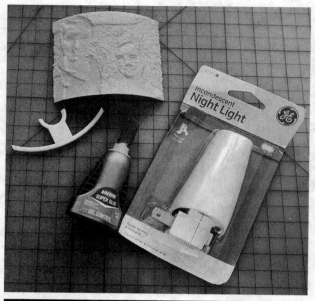

Figure 12-17　The parts.

Figure 12-18　The assembled night-light.

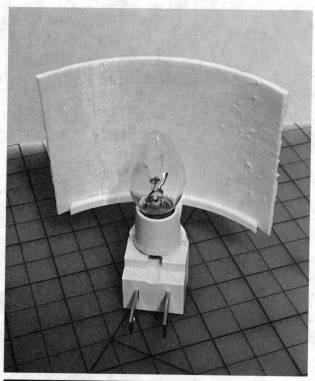

Figure 12-19　The back of the night-light.

Skull-and-Bones Pencil Cup

PROBLEM: An industrial designer wants to display his digital modeling skills with a funky pencil cup. Our cup will be cylindrical in shape with a tapered top, six compartments, and a skull-and-crossbones design modeled from a downloaded SVG file. We'll model it in 123D Design, slice it with Cura, and print it in PLA on a Lulzbot Taz 6 printer.

123D Design is part of Autodesk's now-retired 123D suite of apps. However, the Design app works offline and remains popular due to its ease of use. While you can't download it from Autodesk anymore, Google a third-party download site or ask a 3D printing board community for the .exe file; many makers have saved it. Or just replicate the steps shown here in Fusion 360, as the programs are similar.

Sketch a Circle and Three Lines

Launch Design, click on the Primitive Circle tool (Figure 13-1), and then click the tool onto the work plane (keep the 10-mm spacing default). Before going further, let's display the work plane in a way that makes sketching on it easier. Move the mouse to the View Cube, and click on Top for a plan view of the circle (Figure 13-2).

Now draw lines that will serve as dividers inside the pen cup. Click on the Sketch menu's Polyline tool (Figure 13-3). Click the tool anywhere on the circle to select it. Then click one endpoint at the top and another right under it (Figure 13-4). Click on the green checkmark to exit. Draw two separate horizontal lines the same

Things You'll Need

Description	Source	Cost
Computer and Internet access	Your own or one at a makerspace	Variable
Autodesk account	autodesk.com/	Free
123D Design software	123dapp.com	Free
3D printer and slicing software	Your own, one at a public makerspace, or one at an online service bureau	Variable
Thumb drive (needed only for offsite printing)	Computer or electronics store	< $10
Spool of PLA filament	Amazon, Microcenter, or online vendor	Variable

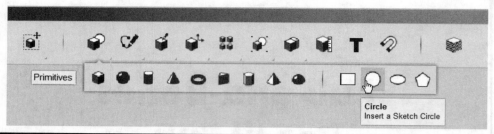

Figure 13-1 Click on the Primitive Circle tool.

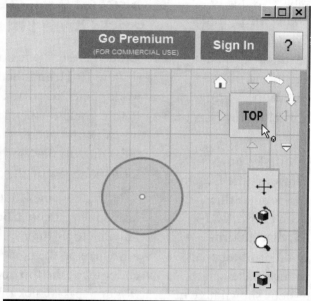

Figure 13-2 Click on the View Cube's Top for a plan view of the circle. The House icon returns the work plane to the default view.

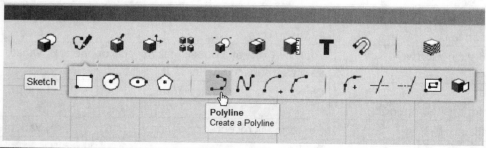

Figure 13-3 The Polyline tool.

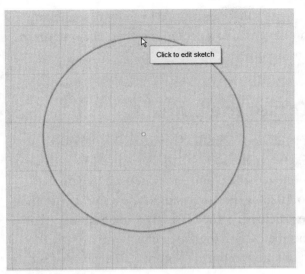

1. Click the circle.

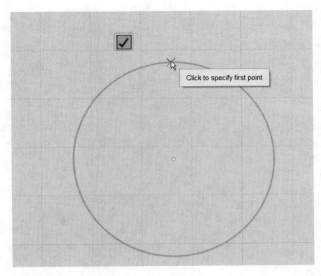

2. Click the first endpoint.

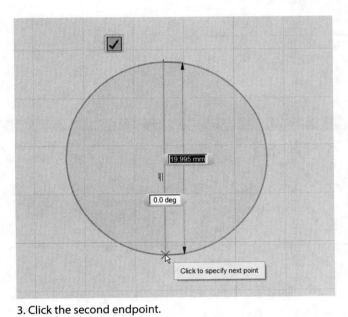

3. Click the second endpoint.

Figure 13-4 Draw a vertical line down the circle's center.

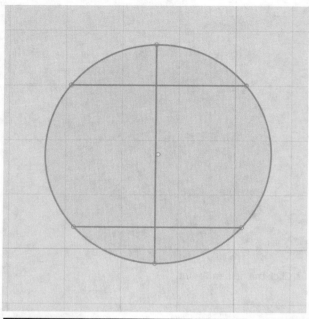

Figure 13-5   Draw two horizontal lines.

way, as shown in Figure 13-5, by first clicking on the circle to select it and then clicking on the endpoints.

Offset the Lines

The divider lines need thickness. Click the Sketch menu's Offset tool (Figure 13-6). Click on a line to offset, and then click where you want to offset it. For precision, type a number in the text field, and the offset line will adjust to that number. Click the green checkmark or hit the ENTER key to finish (Figure 13-7). Then offset the other lines and the circle (Figure 13-8) the same distance.

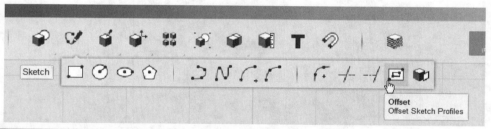

Figure 13-6   The Offset tool.

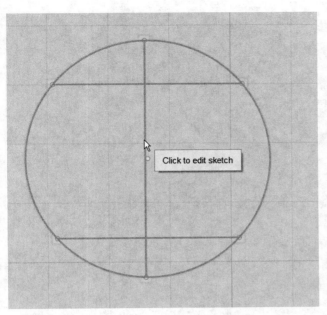

1. Click the line.

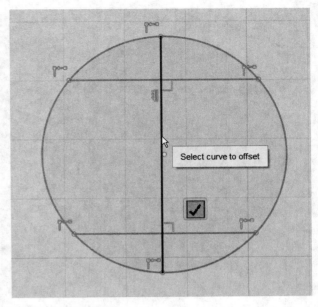

2. Click on the line again.

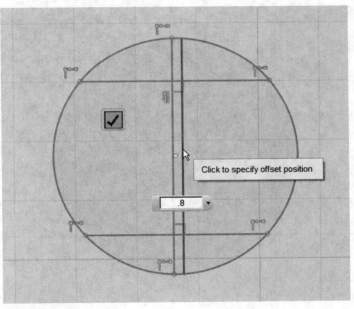

3. Offset the line.

Figure 13-7 Offsetting the vertical line.

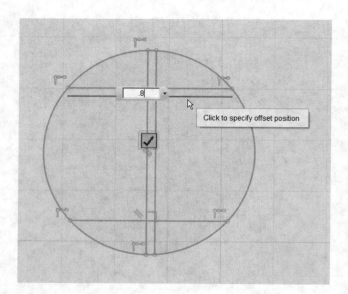
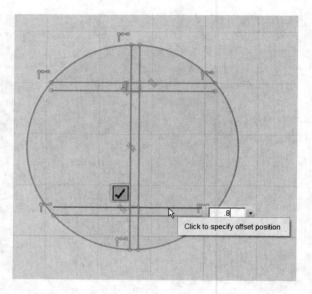
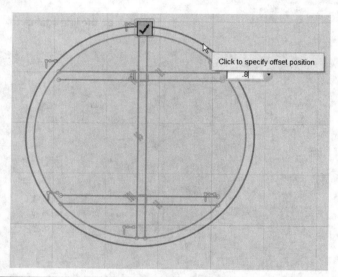

Figure 13-8   Offset all the lines.

Note that two offset lines don't extend all the way to the circle (Figure 13-9). Fix that with the Sketch menu's Extend tool (Figure 13-10). Click on the circle, and then click on a line to extend. The extension will appear in red; hit the ENTER key to finish (Figure 13-11). Extend the other line that needs it, and the cup dividers will be fully sketched.

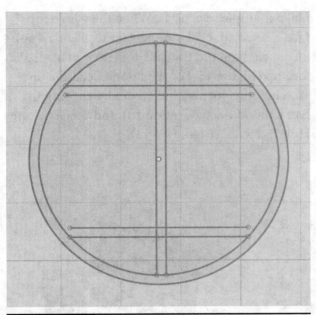

Figure 13-9 The original and offset lines.

Figure 13-10 The Extend tool.

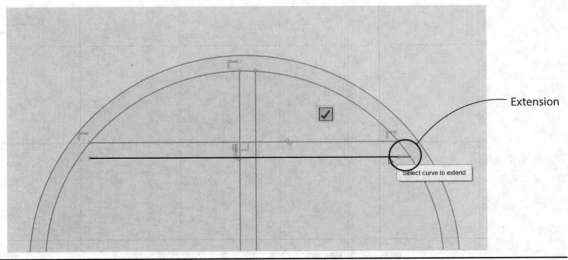

Figure 13-11 Extend the horizontal offset lines so that they touch the circle.

Extrude the Lines

Return the workspace to its default position by hovering the mouse over the View Cube to make the house appear and then clicking on the house. Select the space between all the sketched lines. You'll need to hold the SHIFT key down to make multiple selections. When the whole top face is selected, click on the gear to open a Tools menu, click on the Extrude tool, and extrude the face up (Figure 13-12). You don't need the line sketches anymore, so hide them by clicking the Eye icon on the navigation bar and then clicking Hide Sketches (Figure 13-13).

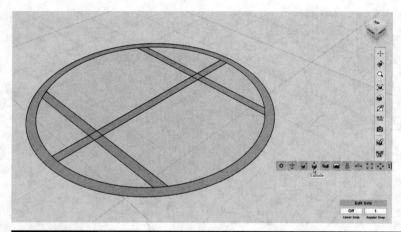
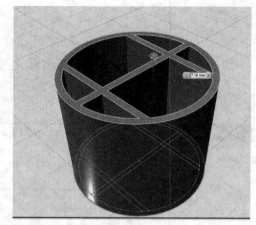

Figure 13-12 Select the space between all the lines, and then click on the Extrude tool.

Figure 13-13 Hide the line sketches.

Find and Import an SVG File

Do a Google images search for "skull and crossbones SVG," and choose a simple black-and-white line-art sketch. I chose the one in Figure 13-14.

Click on Import/SVG as Sketch, and navigate to the file. You'll be shown the options in Figure 13-15. In this case, they will all work the same. Press ENTER, and the file will appear at the corner of the workspace. Drag a window around it, and then, at the top of the screen,

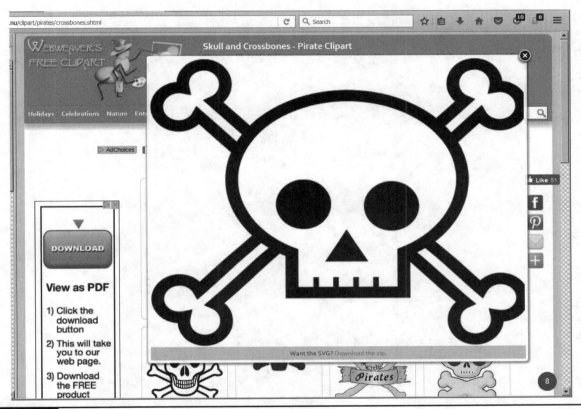

Figure 13-14 An SVG file found with a Google Images search.

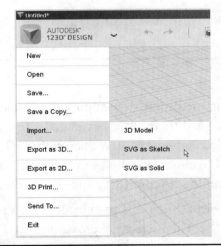

Figure 13-15 Import the SVG file.

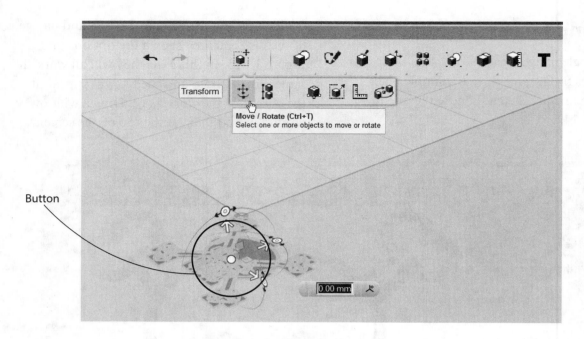

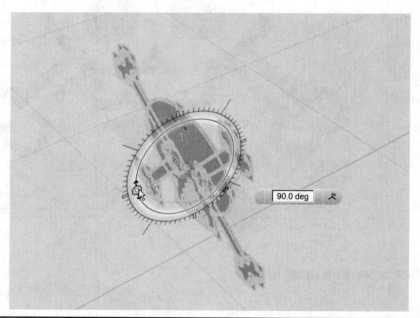

Figure 13-16 Rotate the file 90 degrees.

click the Transform menu's Move tool. Drag the manipulator's top button to rotate the file 90 degrees (Figure 13-16).

Model the SVG File

It's best to make the sketch as simple as possible because complicated ones are difficult to model.

I trimmed some overlapping lines, deleted others (trimmed them away), extended a few, and added a couple of new ones with the Polyline tool to close some holes that resulted from the trimming. Figure 13-17 shows the result.

Select the skull and extrude it forward (Figure 13-18). Then select the crossbones and extrude them forward, but not as far as the skull. Drag

Figure 13-17   The SVG file after trimming, deleting, extending, and adding some lines.

Figure 3-18   Extrude the skull forward.

a crossing window around them both, click the Move icon from the glyph, and position the skull-and-crossbones on the cup (Figure 13-19).

The skull-and-crossbones must be completely inserted into the cup (Figure 13-20). If part is left hanging, such as shown in Figure 13-21, select its face and push/pull it into the cup. You might need to move the whole assembly away from the cup, push/pull it thicker, and then move it back. Part of the assembly will protrude into the cup, but that's okay. We'll fix it later.

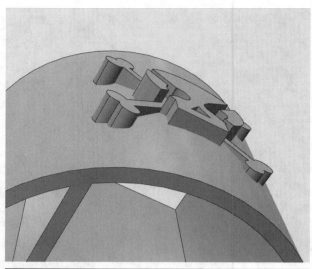

Figure 13-20 The entire model must intersect the cup.

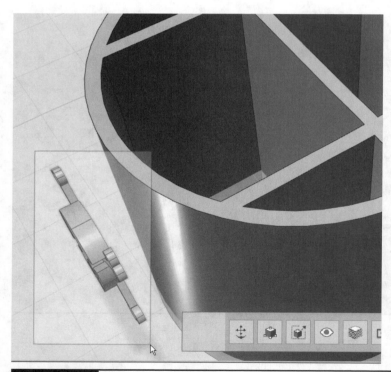

Figure 13-19 Select the model, and move it to the cup.

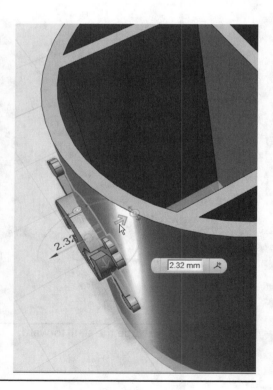

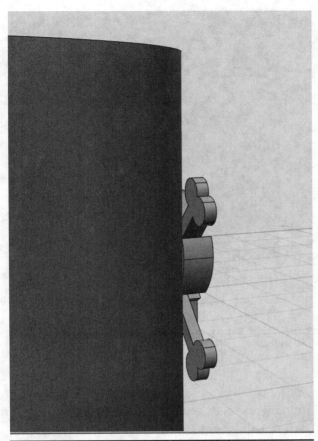

Figure 13-21 If parts don't intersect, push/pull them until they do.

Figure 13-22 Click the grid icon to turn it off.

Add a Bottom to the Cup

Snap a Primitive Circle sketch to the center of the cup's bottom. You might have to turn the sketches back on (Navigation panel/Eye icon/ Show Sketches) to do this. If the grid obscures your view, turn it off by clicking the Eye/Grid icon on the Navigation panel (Figure 13-22). Select the circle, and extrude it up. It will appear red because it is cutting through the cup, but fix that by clicking on the dropdown arrow and choosing New Solid (Figure 13-23).

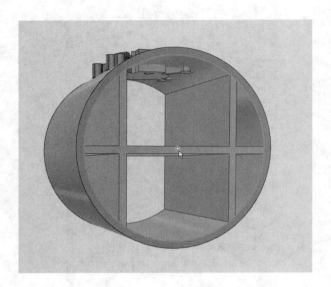

1. Snap a Primitive Circle sketch to the center.

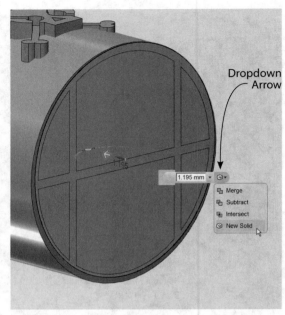

2. Extrude the sketch.

Figure 13-23   Add a bottom to the cup with a Primitive Circle sketch.

Combine All Parts

Everything has to be combined to be 3D printable. Click on the Combine/Merge tool (Figure 13-24). Click on Target Solid/Mesh, and click on the cup. Then click on Source Solid(s)/Mesh(es), and click on the bottom piece. Click on the workspace to finish, and the cup and bottom will be welded together (Figure 13-25). Click on Combine/Merge again, and weld the cup and skull-and-crossbones together in the same way (Figure 13-26).

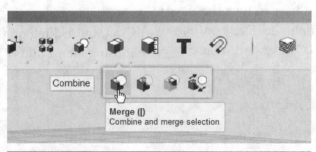

Figure 13-24   Combine the cylinder and the skull-and-crossbones.

Target Solid/Mesh

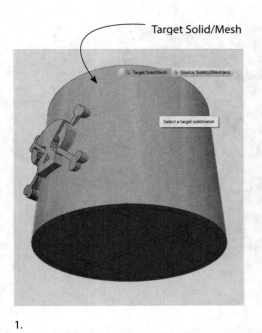

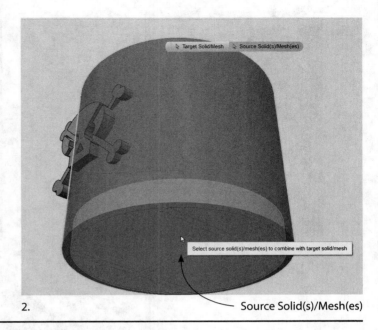

1.

2.

Source Solid(s)/Mesh(es)

Figure 13-25 Combine the cup and bottom.

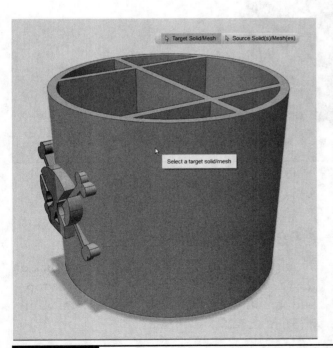

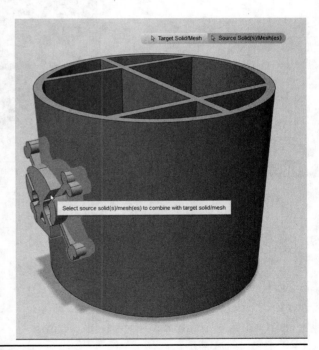

Figure 13-26 Combine the cup and skull-and-crossbones.

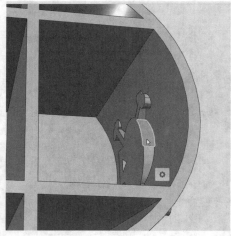

1. Delete.

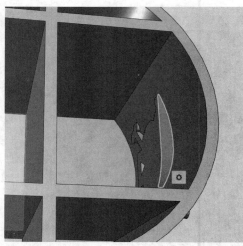

2. Delete more.

Figure 13-27 It took three deletes to completely remove the part protruding inside the cup.

After you weld the cup and skull-and-crossbones together, you can remove the protruding part by selecting it and hitting the DELETE key. The whole part might delete right away, or you might need to hit the DELETE key several times to remove it in portions (Figure 13-27).

Angle the Top

To angle the cup's top, select its top face, click on the gear, and choose Tweak. Then drag the top button until the top is angled the way you want it (Figure 13-28).

Tweak

Button

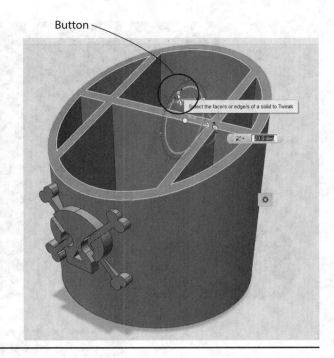

Figure 13-28 Tweak the cup's top.

Adjust Proportions and Chamfer Edges (Optional)

If you don't like the cup's proportions, select it and choose Smart Scale from the glyph. Dimensions will appear around the model, and you can type new ones in the text fields (Figure 13-29). Smart Scale doesn't adjust proportionately, so a number typed on one axis won't affect the others. You can also smooth the edges by selecting them (hold the SHIFT key down to make multiple selections), choosing Fillet from the glyph (Figure 13-30), and dragging the arrow to make the fillet.

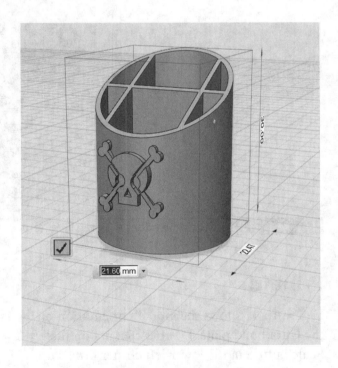

Figure 13-29 Use the Smart Scale tool to change the cup's proportions.

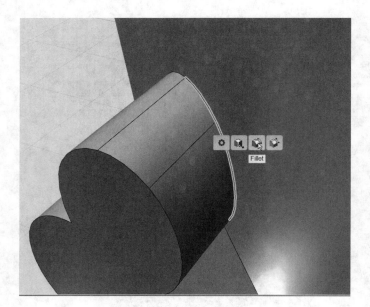

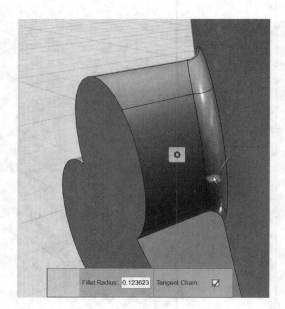

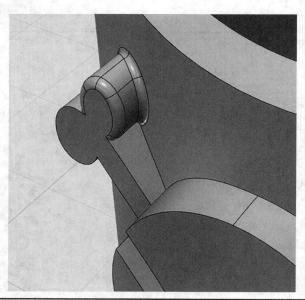

Figure 13-30   Smooth the edges with the Fillet tool.

Print It!

Figure 13-31 shows the model's orientation on the Cura build plate. Double-clicking on the Scale button in the lower-left corner brings up

text fields into which you can adjust the size. The left panel shows some settings. I chose a 15 percent infill. The cup holder (Figure 13-32) was printed on a hot glass build plate covered with PEI and wiped with isopropyl alcohol.

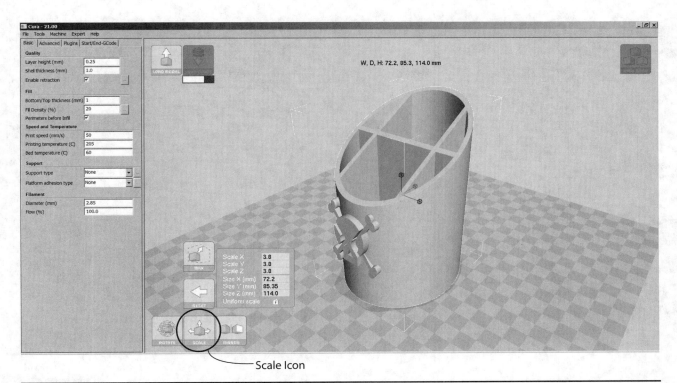

Scale Icon

Figure 13-31 Settings and orientation in Cura.

Figure 13-32 The printed cup holder.

Business Card with QR Code

PROBLEM: An entrepreneur wants some 3D printed cards with a QR code of his website address to give to customers. We'll whip one up in Tinkercad, make it printable in Meshmixer, slice it in MakerBot Desktop, and print it in two colors on a single-extruder MakerBot Rep2.

Point your browser to tinkercad.com, and log in to your Autodesk account. Tinkercad is a Web app. It saves your work automatically and saves it in the cloud. Once logged in, you'll see your dashboard. Click on *Create legacy design* (Figure 14-1), and the Tinkercad workspace will appear (Figure 14-2). At the time of this writing, the beta version is not fully functional and lacks a feature we need, so we'll use the legacy version. The interfaces on both are similar, anyhow.

Things You'll Need

Description	Source	Cost
Computer and Internet access	Your own or one at a makerspace	Variable
Autodesk account	autodesk.com/	Free
Tinkercad Web app	Tinkercad.com	Free
Autodesk Meshmixer software	Meshmixer.com	Free
3D printer and slicing software	Your own, one at a public makerspace, or one at an online service bureau	Variable
Thumb drive (needed only for offsite printing)	Computer or electronics store	< $10
Spool of PLA filament	Amazon, Microcenter, or online vendor	Variable

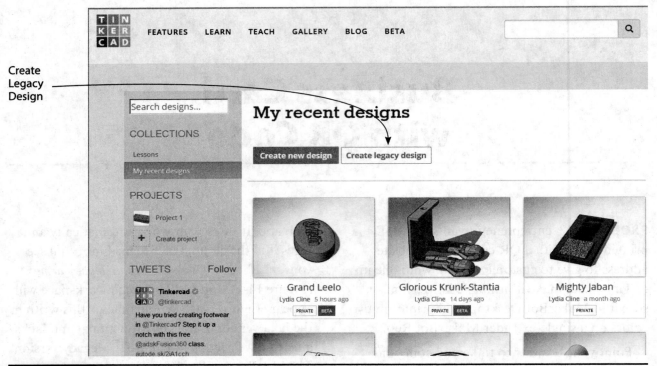

Create
Legacy
Design

Figure 14-1   Click on Create new design.

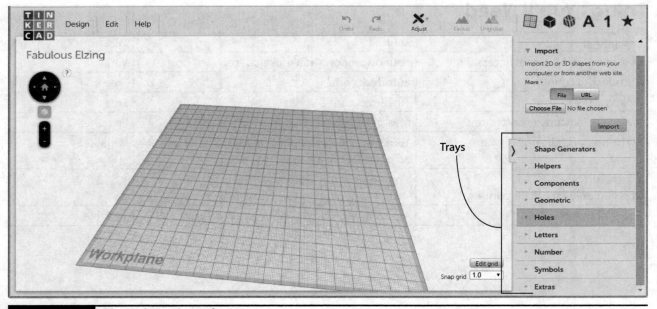

Figure 14-2   The Tinkercad interface.

Model the Card

Note the trays on the right side of the screen. Click on Geometric for a menu of simple forms, and then drag a Box into the work plane (Figure 14-3).

Click on the box to select it, and grips and manipulators will appear. Click on the white button at the top, and drag it down to make the box shorter. Then click on the black side grip and drag to make the box longer. The box should now have a business card shape (Figure 14-4). You can eyeball proportions or type exact numbers into the text fields. Holding the SHIFT key down scales an item proportionately.

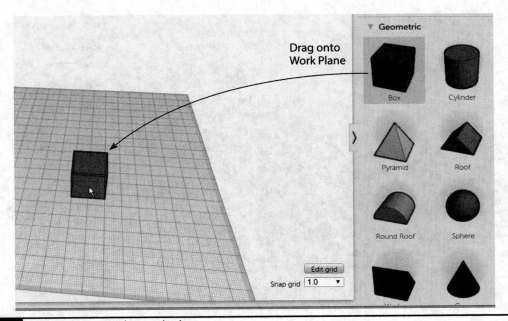

Figure 14-3   Drag a box into the work plane.

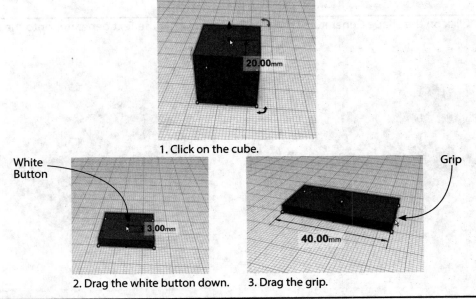

Figure 14-4   Change the shape of the box with grips.

Add Text to the Card

Click on the Shape Generators (Tinkercad) tray. Then click on the Text generator, and drag it onto the card (Figure 14-5). Select the Text generator to make grips appear, and then drag the top grip down to make the text shorter. Drag a center/side grip in to make the text smaller (Figure 14-6).

Now type what you want the card to say into the text field of the Options box that opens with each Shape Generator (Figure 14-7). You can adjust height and font here, too. Simple fonts print best.

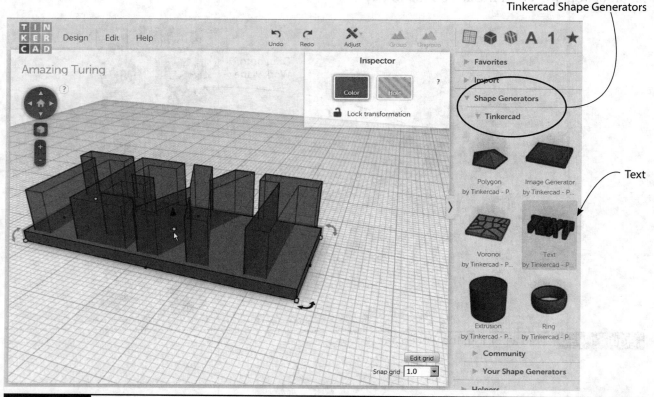

Figure 14-5   Click on the Shape Generators (Tinkercad) tray, and drag the Text generator onto the card.

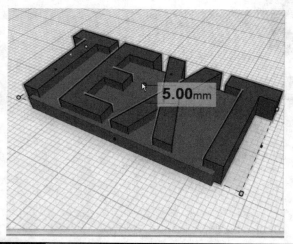
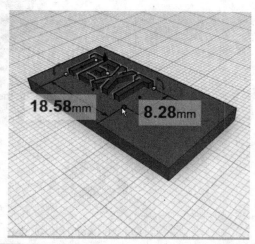

Figure 14-6   Drag the grips to make the text thinner and smaller.

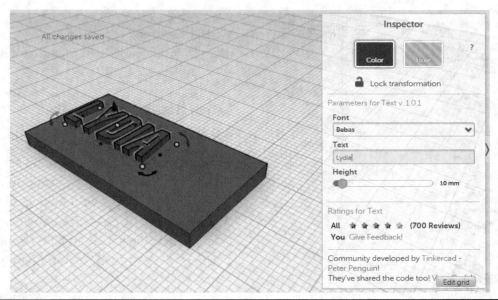

Figure 14-7 Options for the Text generator.

Add a QR Code to the Card

Place a work plane on the card to make putting items on it easier. Click on the Helpers tray and then on Workplane. Drag the work plane onto the card (Figure 14-8).

Now click on the Shape Generators (Community) tray. On page 13 there's a QR code generator (Figure 14-9). Drag it onto the work plane, type a URL into its options box, and then move and size it the same way as the text (Figure 14-10). You can also adjust any

Drag the work plane onto the card.

Figure 14-8 Drag a work plane onto the card.

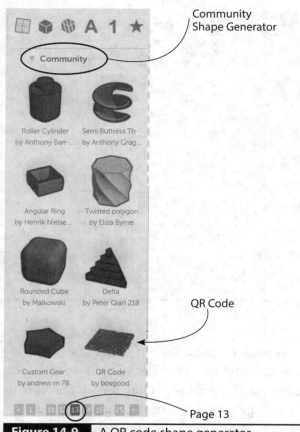

Community Shape Generator

QR Code

Page 13

Figure 14-9   A QR code shape generator.

proportions you want of the card or items on it. When everything looks the way you want, click on the Tinkercad icon in the upper-left screen.

Download the Card as an STL File

Clicking on the Tinkercad icon takes you to your dashboard. Click on a project thumbnail, and a window will appear with download options (Figure 14-11). Click on *Download for 3D Printing*, and then click on STL (Figure 14-12). The file will download to your computer.

Figure 14-10   Enter a URL into the Options box, and move and size the QR generator.

1. Click on Project Thumbnail

2. Click on *Download for 3D Printing*

Figure 14-11 Click on the business card project in the dashboard, and a window will appear with download options.

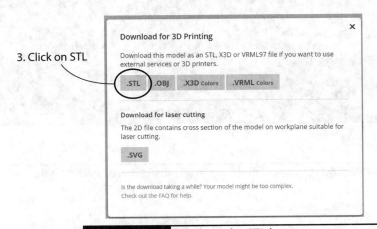

3. Click on STL

Figure 14-12 Click on the STL button.

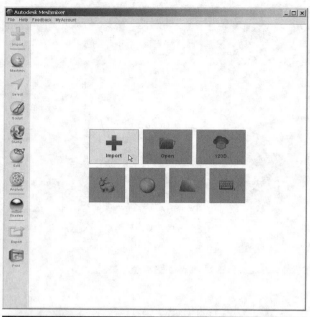

Figure 14-13 Import the file into Meshmixer.

Analyze the File for Printability

Import the STL file into Meshmixer (Figure 14-13). Then click on the Analysis icon and choose Units/Dimensions. The card's dimensions will appear (Figure 14-14), and you can change them by typing a new number in one of the text fields. The other numbers will scale proportionately.

Now click on the Inspector tool that's also found under the Analysis menu. It finds flaws in the file. The file was deeply flawed, and Meshmixer's *Auto Repair All* feature left it looking like Figure 14-15.

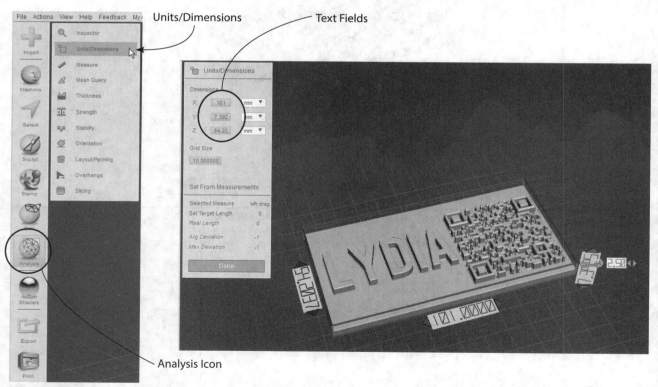

Figure 14-14 Change the file's size by typing new numbers in a Units/Dimensions text field.

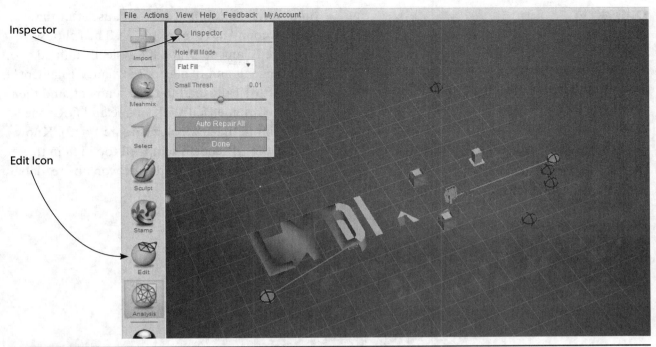

Figure 14-15 Meshmixer's *Auto Repair All* left the file looking like this.

Click on the Edit icon, and choose Make Solid. This fixed the file, and you can finesse the results by moving the Solid Accuracy slider in the Options box all the way to the right (Figure 14-16). Running the Inspector tool again on this file now finds no flaws, so it is ready to print.

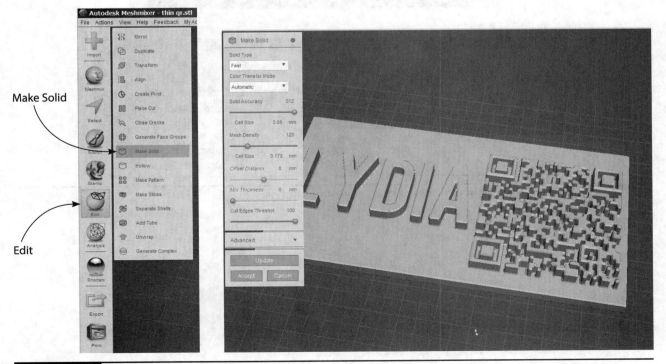

Figure 14-16 Fix the file with the Make Solid tool.

Print It!

Figure 14-17 shows the file's orientation on the build plate and some settings. I chose a 10 percent infill and printed it on an acrylic bed covered in painter's tape wiped with isopropyl alcohol.

To create a two-color card using MakerBot Desktop and one extruder, watch the card under construction and pause it at the appropriate location. Choose Pause after the base is complete (Figure 14-18). Then choose Change Filament/Unload. Physically unload the filament, and then choose Change Filament/Load. Physically load the new filament, and then choose Resume Build. The name and QR code will print in the new color (Figure 14-19). Know that if the QR code is extruded too high in the digital model, the printed code won't be readable with a scanner.

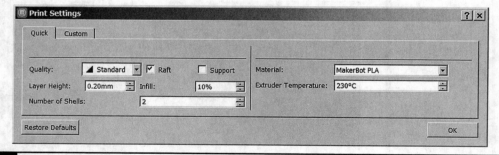

Figure 14-17 The card's orientation on the build plate and settings.

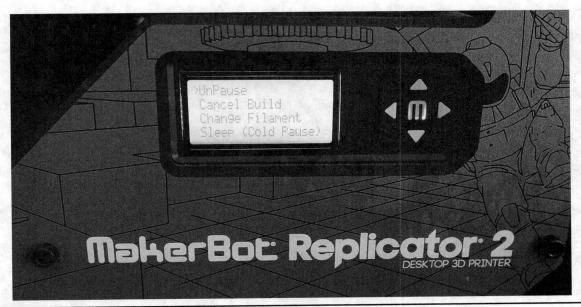

Figure 14-18   Pause the MakerBot either directly or through the MakerBot Desktop slicer.

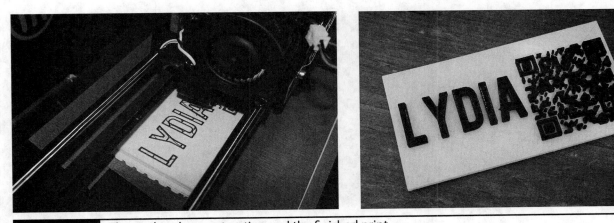

Figure 14-19   The card under construction and the finished print.

From Scan to Trinket Holder

PROBLEM: A community college professor wants to show students how to use a consumer-grade scanner for digital modeling and 3D printing. In this project we'll scan a vase with a Fuel3D Scanify camera, repair and extrude it with Fuel Studio, and turn it into a trinket holder with Meshmixer. Then we'll slice it with MakerBot Desktop and print it with a MakerBot Rep 2.

The Scanify has a dual-laser stereo camera and three xenon flashes that, combined, provide very high-resolution imagery. If you're evaluating similar scanners, this project may offer insight.

What the Scanify Can Do

The Scanify is for small (under 18" tall) items and must be held about 12" away from them. Best results come from colorful, nonglossy items. You can take one photo for a front-facing model; take up to nine photos and stitch them locally for a complete model; or take unlimited photos and stitch them on Fuel3D's cloud. However, the ability to get a model with just one photo is Scanify's advantage over the reality capture technique that requires multiple photos stitched together. This enables you to scan a pet or other subject that won't sit still.

Things You'll Need

Description	Source	Cost
Computer and Internet access	Your own or one at a makerspace	Variable
Autodesk account	autodesk.com/	Free
Autodesk Meshmixer software	Meshmixer.com	Free
Fuel 3D Scanify scanner	scanify.fuel-3d.com/	$1,490
Fuel 3D Studio Plus or Advanced software	scanify.fuel-3d.com/	Subscription $2,000–$2,500/year
3D printer and slicing software	Your own, one at a public makerspace, or one at an online service bureau	Variable
Thumb drive (needed only for offsite printing)	Computer or electronics store	< $10
Spool of PLA filament	Amazon, Microcenter, or online vendor	Variable

What's Needed to Operate the Scanify

Figure 15-1 shows the Scanify. It is handheld and plugs into an electrical outlet and a computer USB port. If you want it to be portable, add a tablet computer and attachments as described on the company website. Studio Advanced software is used in this project (Figure 15-2).

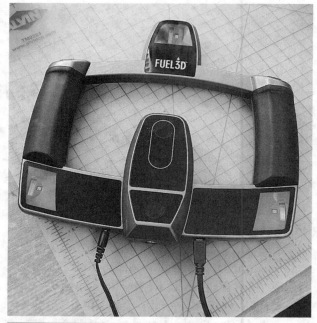

Figure 15-1   The Fuel3D Scanify handheld scanner.

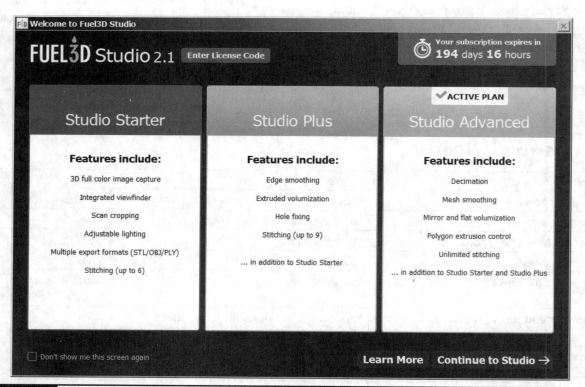

Figure 15-2   Fuel 3D Studio software.

The Scanify comes with scanner targets to place near the subject (Figure 15-3). They rest in handheld or freestanding holders that you print yourself with free files on Sketchfab.com, an STL sharing site (Figure 15-4). You need to make a free account to download them. Links for the holders are located at the end of this project.

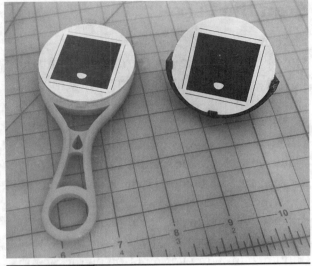

Figure 15-3   Target scanners in 3D printed holders.

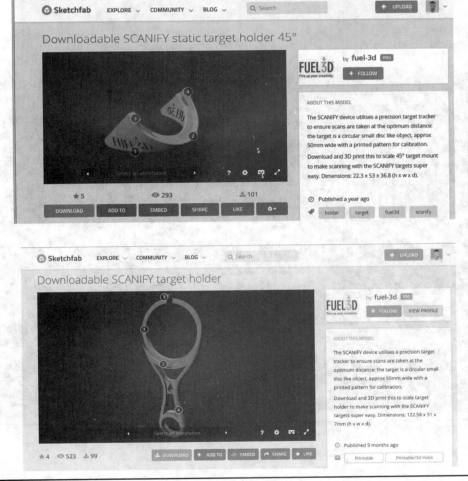

Figure 15-4   Download the holders' STL files at Sketchfab.com.

Scan the Vase

Launch the Studio software. Click on the plus sign in the Active Projects box, and then fill out the dialog box that appears (Figure 15-5). Then click the Viewfinder button. A viewfinder screen appears.

This project's subject is the figurine vase in Figure 15-6. Note the scanner target next to it in the freestanding holder. Hone in on the figurine, and watch the screen for a green dashed circle that goes around the scanner target (Figure 15-7). When the circle appears, either click the Capture button or click a button on the Scanify. I just want one picture of the vase's front. After capturing, the scan appears on the screen; click the green Save button (Figure 15-8).

Figure 15-6 The vase on a chair. The target scanner is next to it, propped up in a 3D printed holder.

Plus Sign

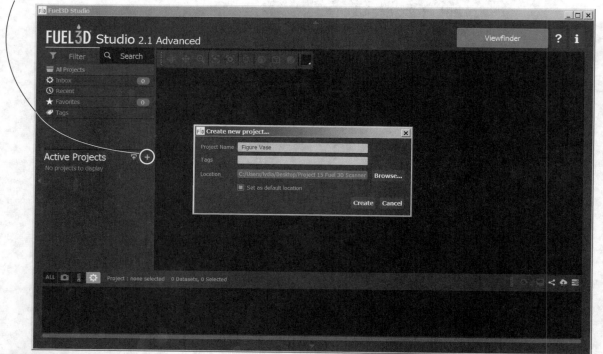

Figure 15-5 Start a new project.

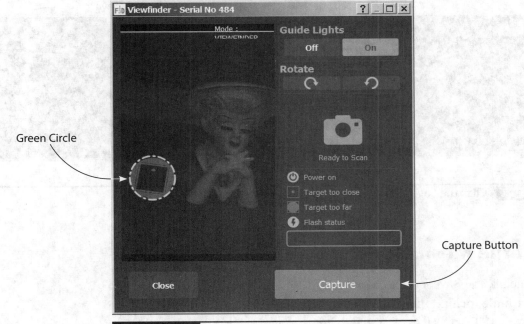

Figure 15-7 Move the Scanify around until a green dashed circle appears around the target scanner on the screen. Then click the Capture button.

Figure 15-8 Save the scan.

Tip: Diffused, even lighting with no glare or shadows is needed for a successful scan. Bright, matte colors scan best. Glossy or reflective surfaces do not scan well.

A thumbnail of the saved scan will appear at the bottom of the screen. Figure 15-9 shows thumbnails of three photos I took, not for stitching but simply for a choice of the best one. At this point I can either save the scan as a mesh model by clicking on the Save icon (Figure 15-9) or work on it in the Studio software. Let's discuss saving it first because if you don't have the pay version of the Studio software, you'll need to save and export it into another program such as Meshmixer to work on.

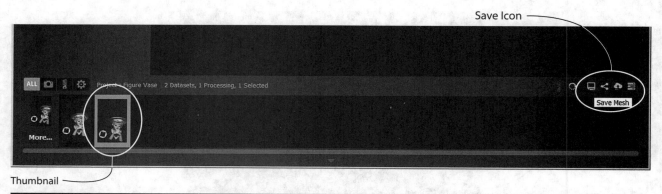

Save Icon

Thumbnail

Figure 15-9 Click the Save icon to save the scan as a mesh model.

Export the Mesh Model

When you click the Save icon, a dialog box appears with file format options. You can save the mesh as an STL, OBJ, or PLY file format (Figure 15-10). If you save it as an OBJ file, an MTL file that contains color information will save with it (Figure 15-11). Keep the two files together in their own folder because when you import the OBJ file into another program, that

FigureVase2.mtl   FigureVase2.obj

Figure 15-11 The OBJ file saves with an MTL file containing color information.

program will read the accompanying MTL file at the same time.

Figure 15-12 shows the OBJ file imported into Meshmixer. You can see how high its resolution is. Figure 15-13 shows the same file decimated (reduced number of polygons) with the Edit/Reduce tool. If you can't see the mesh, press and release the w key on the keyboard.

The file needs to have holes repaired and be extruded to make it 3D printable. This can be a lot of work in Meshmixer, so let's abandon this file now and return to the scan in the Studio software.

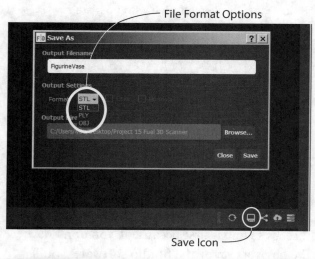

File Format Options

Save Icon

Figure 15-10 Name and export the mesh model as an STL, OBJ, or PLY file.

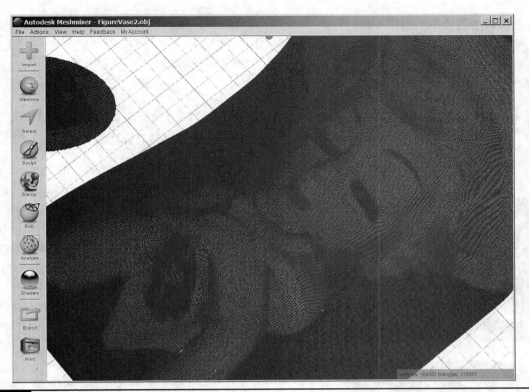

Figure 15-12   The OBJ file imported into Meshmixer.

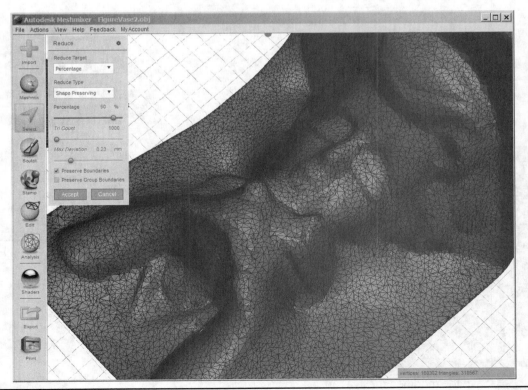

Figure 15-13   The OBJ file with some decimation.

Fix the Mesh Model with Studio Advanced

In the upper-left corner of the Studio screen are menu icons; click on the last one for remeshing tools (Figure 15-14). You can fill holes, decimate, smooth, and volumize (extrude) the mesh by clicking each tool. When finished, click on the Save icon, and choose STL, OBJ, or PLY. We'll save it as an STL file (Figure 15-15).

Import the Vase STL File into Meshmixer

We'll use Meshmixer to cut the bottom of the vase for a smooth surface. Launch it and import the STL file (Figure 15-16). Note that the scanner target also got extruded in Studio and imported. Click the Select tool and then click a selection spot onto the target. Next, click Modify/Expand to Connected to select the whole target. Press and release the DELETE key to remove it (Figure 15-17).

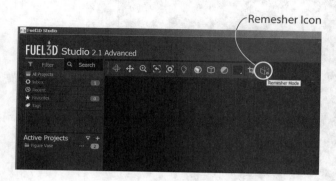

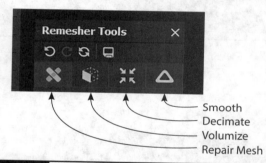

Figure 15-14 Click on the Remesher Mode icon to access remesher tools.

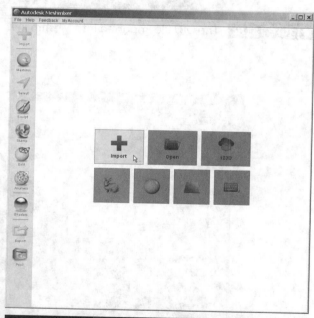

Figure 15-16 Launch Meshmixer and import the STL file.

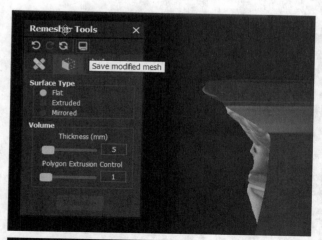

Figure 15-15 Save the modified mesh as an STL file.

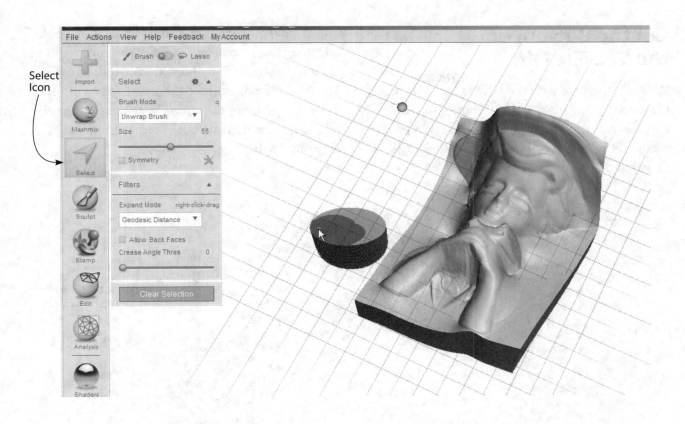

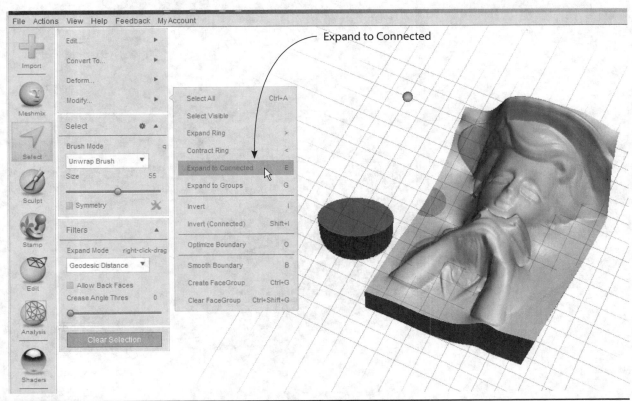

Figure 15-17   Select and delete the unneeded item.

Cut the Bottom of the STL File Off

We'll cut the vase with the Plane Cut tool (Figure 15-18). Click on Edit/Plane Cut. A cutting plane and manipulator appear on the model. Drag a curved line to rotate the cutting plane, and hold the mouse directly over the radial marks while doing so to rotate it in whole degrees. Drag an arrow to place the cutting plane at a specific location. The thick purple arrow points in the direction of the part to be cut off, so if it's pointing the wrong way, use the curved line to rotate it. Then click Accept to finish (Figure 15-19).

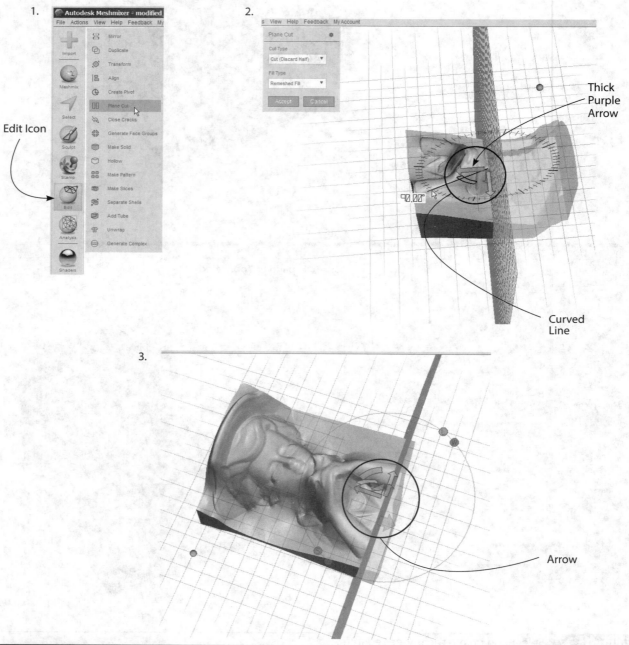

Figure 15-18   Position a cutting plane.

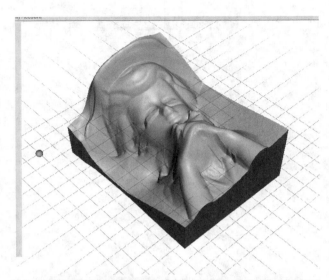

Press and release the T key on the keyboard to bring up the Move tool, and then use a curved line to position the file so that it stands up (this makes it easier to work with). You can also drag the file anywhere on the work plane by dragging the arrows. Click Accept to finish the move (Figure 15-20).

Figure 15-19 The file now has a straight, smooth bottom.

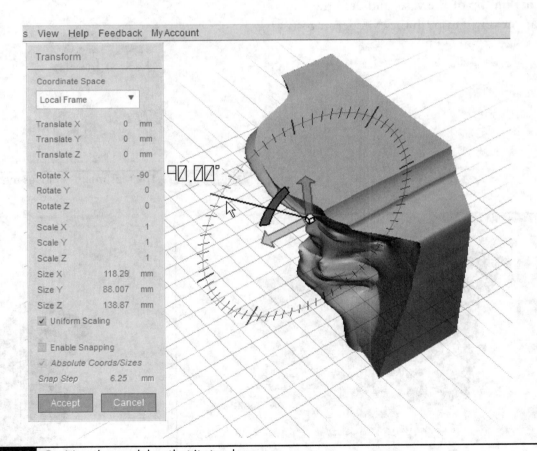

Figure 15-20 Position the model so that it stands up.

Cut a Hole in the Top of the STL File

Click on the Meshmix icon and then on the Primitives menu (Figure 15-21). Drag a cylinder into the work plane, but not onto the vase; instead, just drop it in space (Figure 15-22). Note that two models now show up in the Object Browser box (if the Object Browser box isn't visible, click the View menu at the top of the screen, and then click Show Object Browser). The cylinder is highlighted in this box, meaning that it's active. Press and release the т key on the keyboard to make the Move tool appear over it. Then drag the Move tool's white box to make the cylinder smaller, drag an arrow to position the cylinder on top of the vase, and drag the vertical arrow down to push the cylinder into the vase.

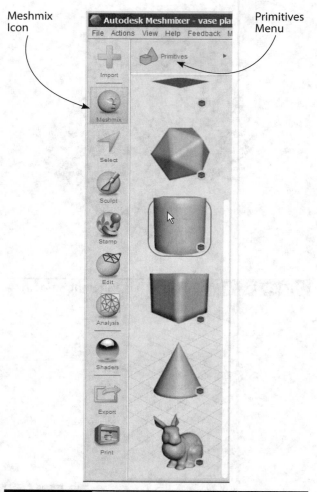

Meshmix Icon

Primitives Menu

Figure 15-21 The Meshmix/Primitives menu.

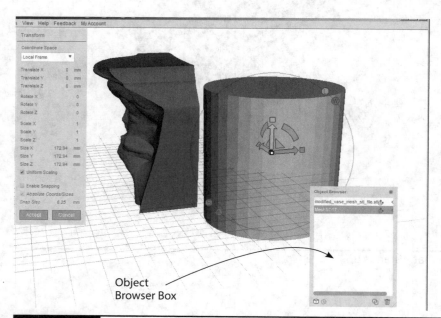

Object Browser Box

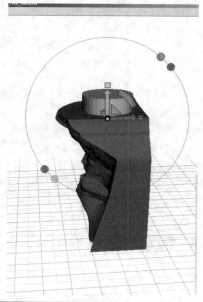

Figure 15-22 Drag a cylinder into the work plane, scale it down, and move it on top of the vase.

You can make the vase transparent by dragging the transparency shader onto it (Figure 15-23). This helps you to eyeball the proper depth to push the cylinder. Drag the opaque shader onto the vase when you don't need transparency anymore.

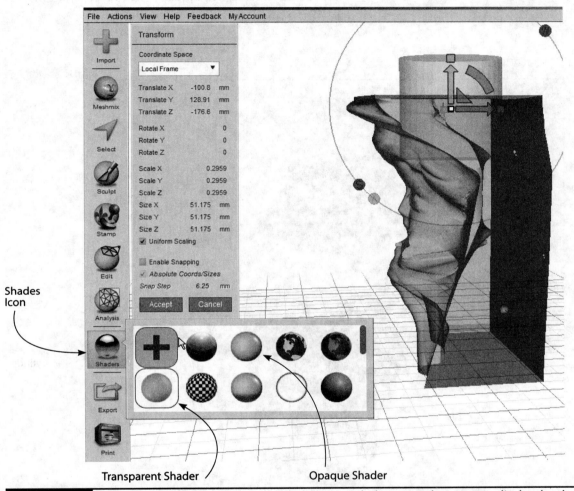

Shades Icon

Transparent Shader

Opaque Shader

Figure 15-23 Drag the transparency shader onto the vase to help gauge the proper cylinder depth.

Click on the vase entry in the Object Browser. Then hold the SHIFT key down and click on the cylinder entry. Both entries are now selected, and a new Edit menu with Boolean operations appears (Figure 15-24). The order of selection is important; clicking on the vase entry first means that the cylinder will be subtracted from it.

Click Boolean difference on the menu, and the cylinder will subtract from the vase. An Options box will appear; click Accept to finish the operation (Figure 15-25).

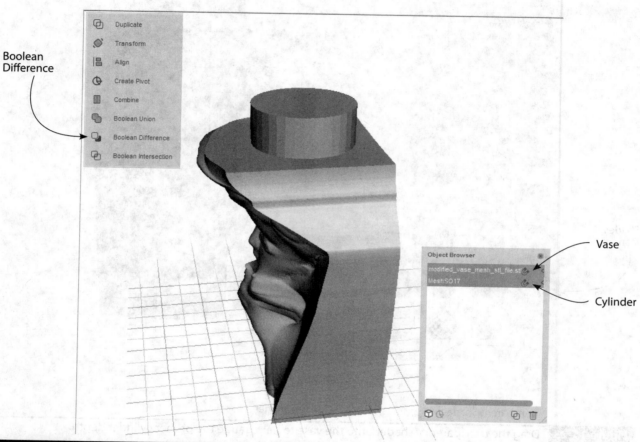

Figure 15-24   Select the vase entry in the Object Browser first, hold the SHIFT key down, and select the cylinder entry second.

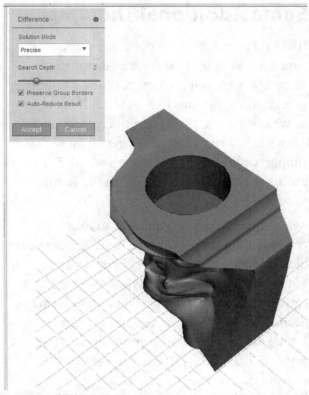

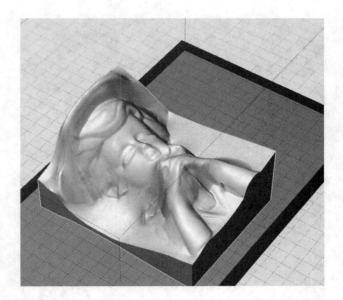

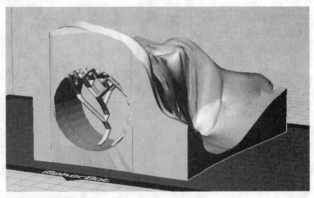

Figure 15-25 After subtracting the cylinder from the vase, click Accept. We now have a trinket holder.

Figure 15-26 The model's orientation on the build plate.

Print It!

Figure 15-26 shows the trinket holder's orientation on the build plate. Only the hole needed support. Figure 15-27 shows settings, and Figure 15-28 is the print. It was printed on painter's tape wiped with isopropyl alcohol.

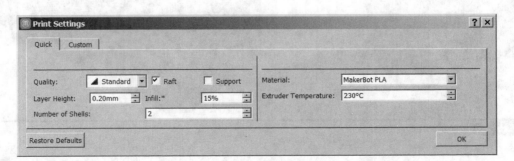

Figure 15-27 The print's settings.

Figure 15-28   The printed trinket holder.

Some Additional Thoughts

It's best to analyze build plate orientation yourself before using software-generated orientation and support suggestions. Meshmixer generated the options in Figure 15-29, neither of which is good; the first option would likely damage the print after removing all those supports, and the second option would fail during printing due to weight on the skimpy supports.

I laid the model flat, and Meshmixer generated the supports in the hole. The supports didn't develop successfully (Figure 15-30), but the hole printed well anyhow. Sometimes software generates supports that really aren't needed. As an aside, the purple filament in Figure 15-30 is from a 3D printing pen. The print split from the raft early in the process, causing the right side to begin curling up. Welding the print back to the raft with the printing pen saved it.

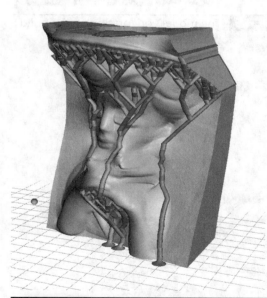

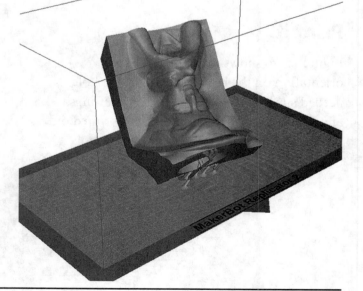

Figure 15-29   Orientation options.

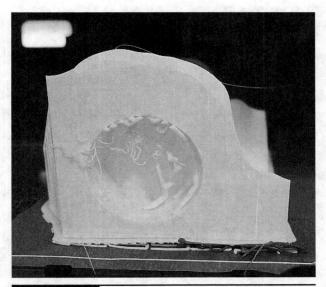

Figure 15-30 Meshmixer's supports didn't develop well in this print, but they weren't needed, anyhow.

Further Resources

- Static target holder: https://sketchfab.com/ models/c08466083b514caf878a4614836c2 402?utm_source=triggered-emails&utm _medium=email&utm_campaign=model -liked#

- Handheld target holder: https://sketchfab .com/models/9f191204dfb0498091fa226ae 381d62b

- Lydia's YouTube review of the Scanify: youtube.com/watch?v=y8C2fz7H0MU

Phone Stand

PROBLEM: A student wants a fun way to prop his phone up. In this project we'll download an STL file from Thingiverse and add a base to it in Fusion 360. Then we'll slice it with Simplify3D and print it in PLA on a Gmax 1.5XT+ printer.

Use the search engine at yeggi.com to find an upright model that's roughly the width of your phone (Figure 16-1). Alternatively, find two upright, thin models. I used the model at this link: www.thingiverse.com/thing:504230.

Import the STL File into Fusion 360

Launch Fusion 360, and then click Insert/Insert Mesh. A navigation browser will appear; find the STL file, and press ENTER (Figure 16-2).

The STL file enters with the Move widget over it. Drag a button to rotate it upright (Figure 16-3), and then click OK.

Things You'll Need

Description	Source	Cost
Computer and Internet access	Your own or one at a makerspace	Variable
Autodesk account	autodesk.com/	Free
Autodesk Fusion 360 software	autodesk.com/products/fusion-360/overview	Free or subscription cost
3D printer and slicing software	Your own, one at a public makerspace, or one at an online service bureau	Variable
Thumb drive (needed only for offsite printing)	Computer or electronics store	< $10
Spool of PLA filament	Amazon, Microcenter, or online vendor	Variable

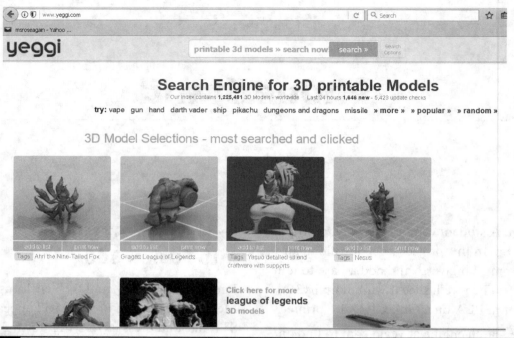

Figure 16-1   STL file from Thingiverse.

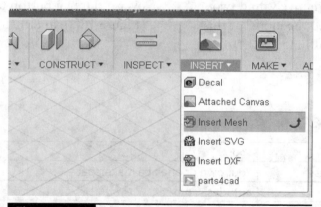

Figure 16-2   Import the STL file.

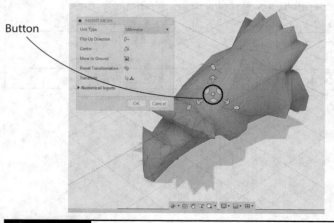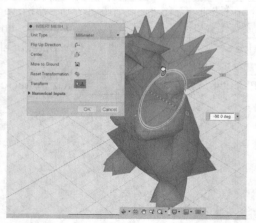

Button

Figure 16-3   Rotate the STL file upright.

Model a Base

The model needs a base to stand on. To make drawing it easier, click the Top view on the View Cube in the upper-right corner. Then click on the dropdown arrow that appears when you hover the mouse over the View Cube, and click Orthographic (Figure 16-4).

Click Sketch/Rectangle/2-Point Rectangle (Figure 16-5). Click it once on the grid, click a second time to choose the first endpoint, and click a third time to choose the last endpoint (Figure 16-6). Type the exact dimensions in the text fields; I just eyeballed proportions. The file can be scaled in the slicer later. Next, while still

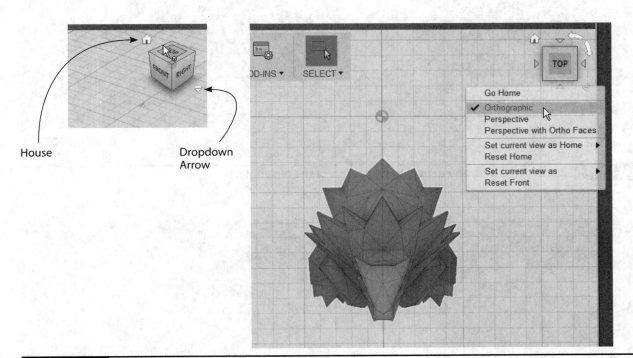

House   Dropdown Arrow

Figure 16-4   View the model top down and orthographically.

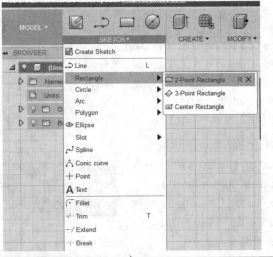

Figure 16-5   Click Sketch/Rectangle/2-Point Rectangle.

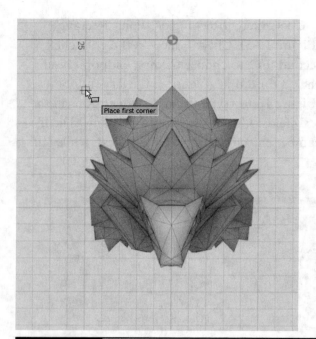

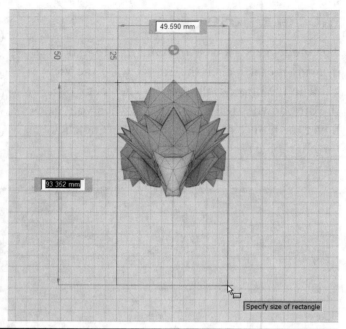

Figure 16-6   Click on the grid, and then click two points of the rectangle.

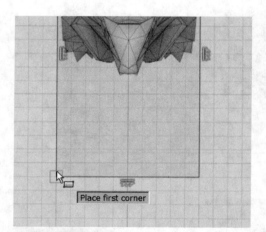

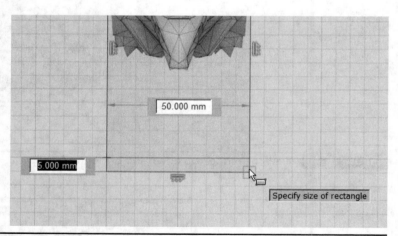

Figure 16-7   Draw a rectangle to serve as a phone stop.

in the Sketch command, draw a thin rectangle to serve as a phone stop (Figure 16-7).

Return to a 3D position either by clicking the house that appears when hovering the mouse over the View Cube or by clicking the Orbit icon at the bottom of the screen (Figure 16-8). You can also orbit by pressing and holding the SHIFT key first and then pressing the mouse's scroll wheel down second. Hit the ESC key to get out of Orbit mode.

Figure 16-8   The Orbit tool.

The rectangles aren't level with the model's feet, so drag a selection window around them, right-click, and choose Move from the option tree (Figure 16-9). Click the right side of the View Cube to make aligning the rectangles with the feet easier.

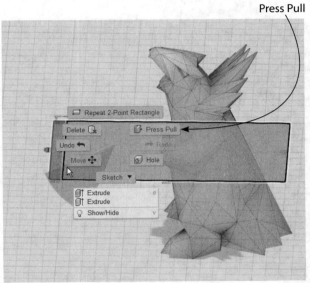

1. Right-click and choose Press Pull.

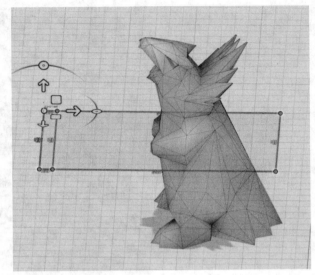

2. Move down.

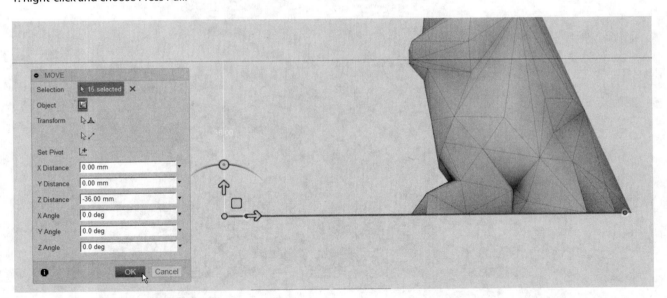

3. Align rectangles with bottom of model.

Figure 16-9   Align the rectangles with the model's feet.

Select the underside of the base, right-click, choose Press Pull, extrude it down, and click OK (Figure 16-10). Then select the top of the stop, extrude it up, and click OK (Figure 16-11). If

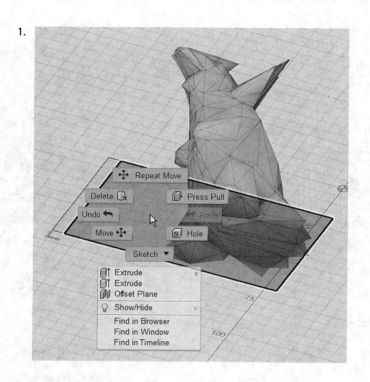

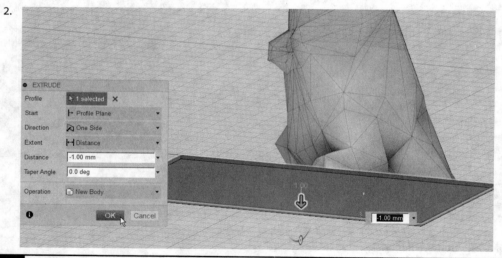

Figure 16-10 Extrude the large rectangle down.

needed, adjust the sides to align with the STL file. To make this easier, orient the phone rest so that it faces you by selecting its front face and clicking on the Look At icon (Figure 16-12).

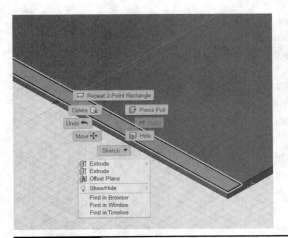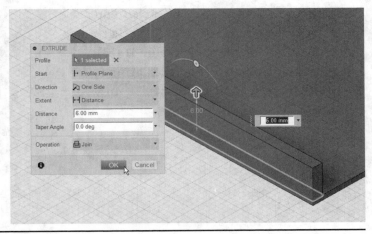

Figure 16-11 Extrude the small rectangle up.

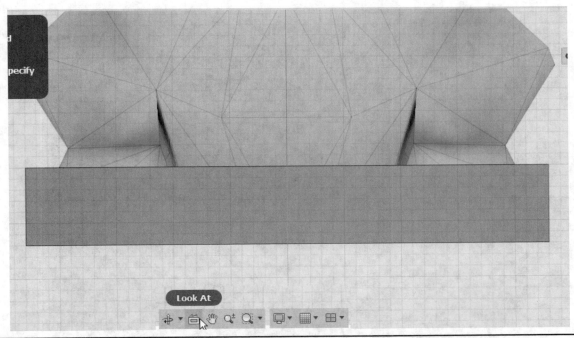

Figure 16-12 Orient the phone rest to face you by selecting its front face and clicking on the Look At icon.

Make a Hole in the Phone Stop

Now we need a hole through which to place a charging cord. Click on Sketch/Circle/Center Diameter Circle (Figure 16-13). Then click on the center of the phone stop (Figure 16-13). Drag the mouse outward, and click again to set the circle's diameter (Figure 16-14).

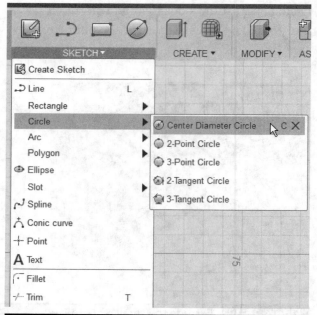

Figure 16-13 Click on the Sketch/Circle/Center Diameter Circle tool.

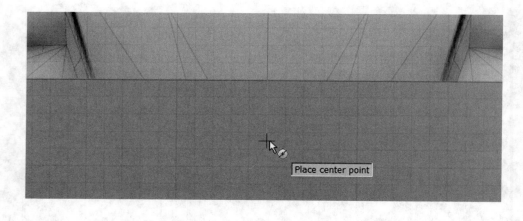

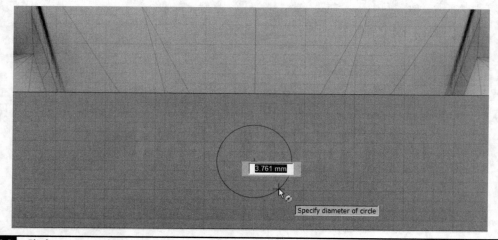

Figure 16-14 Click twice to sketch the circle.

Tip: You can find the center by using the grid or by drawing guidelines first with the Sketch/Line tool and hovering that tool over the face's edges until the triangle symbol that indicates a midpoint appears.

Select the circle, right-click, and choose Press Pull. Extrude it through the phone rest, and then click OK (Figure 16-15). This turns the circle into a hole.

Convert the STL File to a Solid

Converting the STL file to a solid will enable us to combine it with the base, which is necessary to make it 3D printable. Right-click on the file's name in the Browser box; in this case, it's called "Untitled" (Figure 16-16). Choose Do Not Capture Design History. A warning dialog will appear; click Continue.

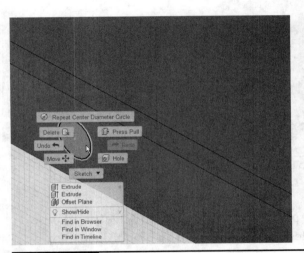
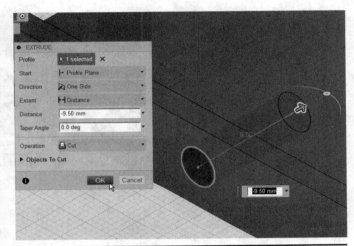

Figure 16-15 Turn the circle into a hole.

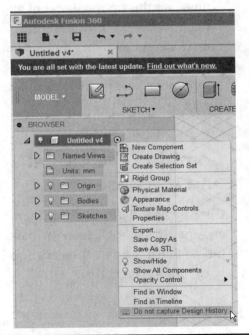
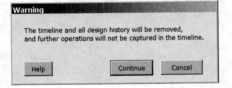

Figure 16-16 Turn off Design History.

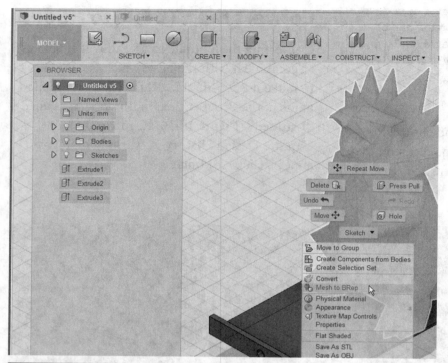

Figure 16-17   Right-click on the STL file, and choose Mesh to BRep.

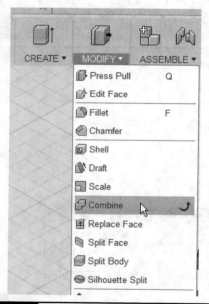

Figure 16-18   Click on Modify/Combine.

Now right-click on the STL file and choose Mesh to BRep (Figure 16-17). A dialog box appears that says that one mesh is selected. Click OK, and the STL file will convert to a solid. Know that this option does not appear until Design History is turned off. Furthermore, it only works on STL files that are perfect, that is, have no holes or other flaws in the mesh. If a conversion isn't possible, you'll get a message saying so.

Combine the Model and the Base

Click on Modify/Combine (Figure 16-18). Then click on the body, click on the base, and click OK to finish the operation (Figure 16-19). Both parts will weld together.

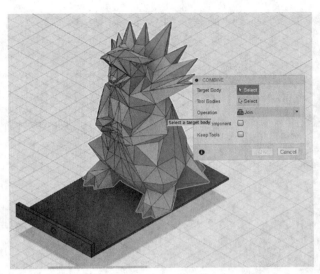

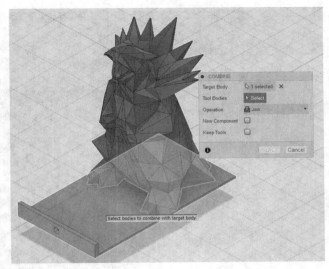

1. Select Typhlosion.

2. Select the base.

Click on the body, then on the base, and then on OK.

Export the Phone Holder as an STL File

Right-click on the name in the Browser box and choose Save as STL file (Figure 16-20). An options box will appear; accept the defaults, and click OK.

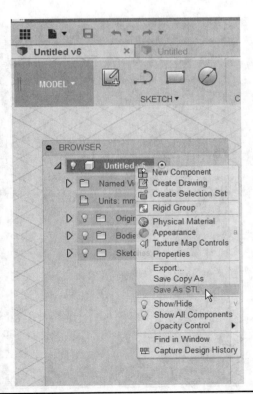

Save as STL file, and accept the defaults.

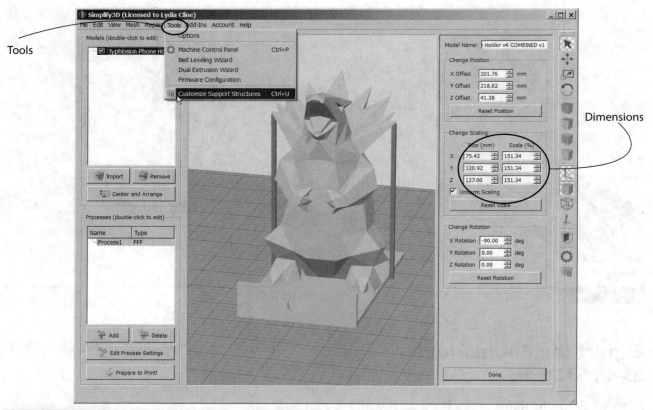

Tools

Dimensions

Figure 16-21 The model inside Simplify3D. The vertical posts are supports.

Print It!

Figure 16-21 shows the model's orientation on the Simplify3D build plate. Double-click on it to bring up dimension text fields. You can accept the existing dimensions or type new ones. Click on Tools/Customize Support Structures to generate supports. The slicer automatically generated two vertical posts, which can be modified. I chose a 15 percent infill.

The phone stand (Figure 16-22) was printed on a cold aluminum build plate covered with painter's tape wiped with isopropyl alcohol.

Figure 16-22 The printed phone stand and supports.

Cage Pendant with a Bead Inside

PROBLEM: A jewelry designer wants to prototype an idea for a necklace pendant. We'll model it in Fusion 360, slice it with Simplify3D, and print it in Colorfabb Bronzefill on a Gmax 1.5XT+ printer.

Enhance a print by adding something to it while under construction. We're going to model a pendant that looks like a cage and add a bead to it. This strategy can also be used to add sand or other filler to make a weighted object.

Make a Torus

Launch Fusion 360. On the menu at the top of the screen, click on Create/Torus (Figure 17-1). Then click on the horizontal plane, click on the origin for the torus's center point, and type *25.4* for the inner diameter (Figure 17-2).

Things You'll Need

Description	Source	Cost
Computer and Internet access	Your own or one at a makerspace	Variable
Autodesk account	autodesk.com/	Free
Autodesk Fusion 360 software	autodesk.com/products/fusion-360/overview	Free or subscription cost
3D printer and slicing software	Your own, one at a public makerspace, or one at an online service bureau	Variable
Thumb drive (needed only for offsite printing)	Computer or electronics store	< $10
Spool of PLA filament	Amazon, Microcenter, or online vendor	Variable

Figure 17-1 On the menu at the top of the screen, click on Create/Torus.

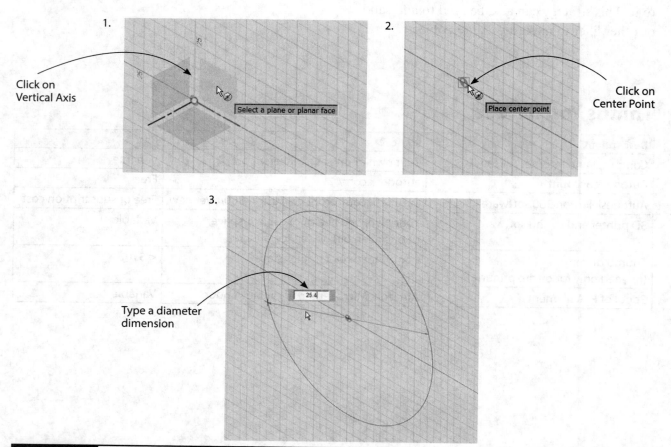

Figure 17-2 Place a torus on the horizontal plane.

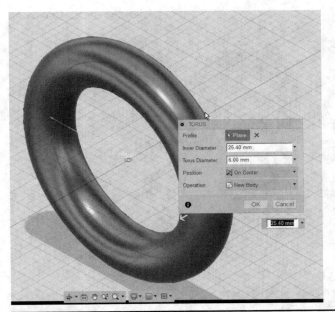

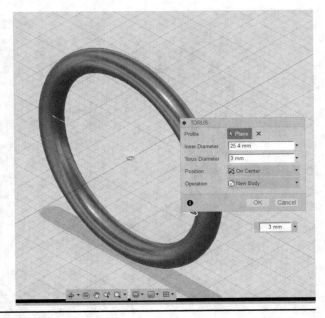

Figure 17-3   Make the torus thinner.

A torus with the diameter specified will appear, and it has a default of 6 mm for the outer diameter, which is shown as the "torus diameter" in the dialog box. Type *3* in the Torus Diameter box to make the torus thinner (Figure 17-3).

Copy and Rotate the Torus

Now make multiple rotated copies to form a cage. Click on Create/Pattern/Circular Pattern (Figure 17-4). Click on the torus to select it, click on the vertical axis, and then type the number of copies you want (Figure 17-5). I typed *7*, and Figure 17-6 shows the result.

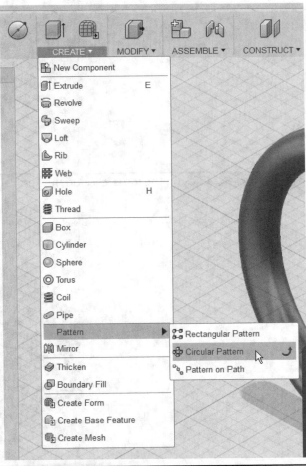

Figure 17-4   Click on Create/Pattern/Circular Pattern.

Select the Torus

Select the Vertical Origin Line

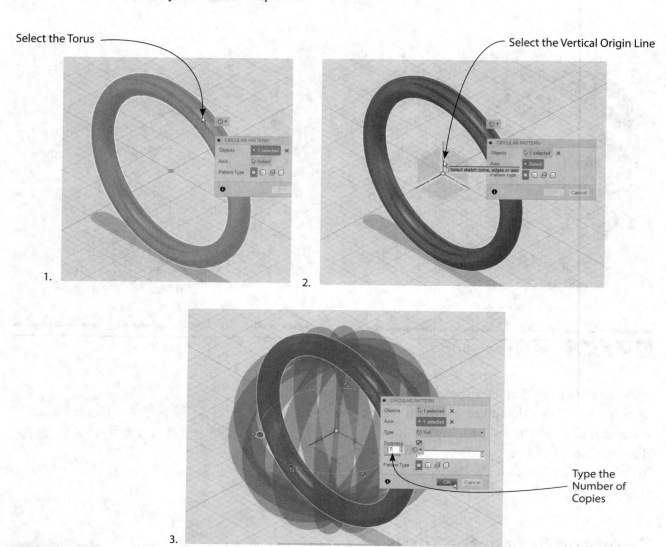

Type the
Number of
Copies

Figure 17-5 Click on the torus, click on the vertical axis, and type the number of copies.

Figure 17-6 The cage pendant.

Make a Loop

The pendant needs a loop on top for attaching a chain. Make a torus next to the cage (Figure 17-7). Click on the horizontal plane, click a center point, and type *15* for a 15-mm interior diameter. Then type *3* to make a 3-mm torus outer diameter.

We need to move the torus. To make positioning it easier, hover the mouse over the View Cube so that the dropdown arrow appears.

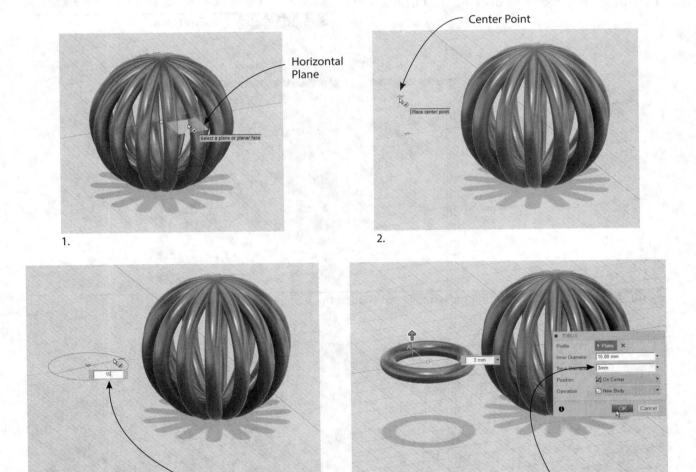

1.

2.

3. Type diameter

4. Type outer diameter

Figure 17-7 Make a torus to serve as a loop.

Click on that arrow, and choose Orthographic (Figure 17-8). Then click on the cube's Front view. Now right-click on the torus, choose Move, rotate it 90 degrees, and then drag the arrows to position it on top of the cage (Figure 17-9). Insert it a bit into the cage, but not enough so that it protrudes through it.

Return to a 3D position by either clicking the house that appears when hovering the mouse over the View Cube or by clicking the Orbit icon

at the bottom of the screen (Figure 17-10). You can also orbit by holding the SHIFT key down first and then holding the mouse's scroll wheel down. Hit the ESC key to get out of Orbit mode.

Figure 17-10 The Orbit tool.

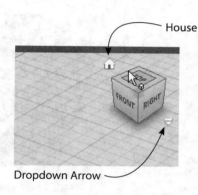

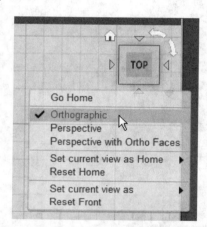

Figure 17-8 Choose Orthographic to make positioning the loop on top of the torus easier.

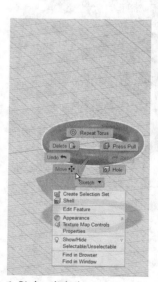

1. Right-click the torus and choose Move.

2. Drag the button.

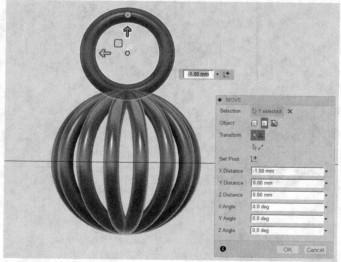

3. Place the loop on the top of the cage.

Figure 17-9 Rotate the torus, and move it on top of the cage.

Combine the Parts and Export as an STL File

We need to weld the loop and all the cage parts together now, a necessary step for this model to be 3D printable. At the top of the screen, click on Modify/Combine (Figure 17-11). Click on the loop, click on Join/Join, drag a selection window around the cage, and then click OK (Figure 17-12). All the parts will combine.

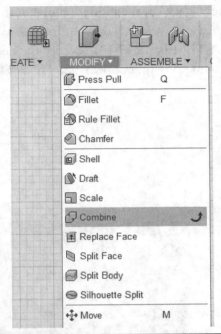

Figure 17-11 Click on Modify/Combine.

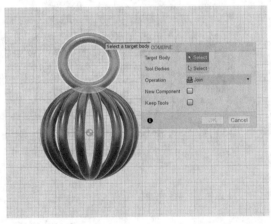

1. Select the loop.

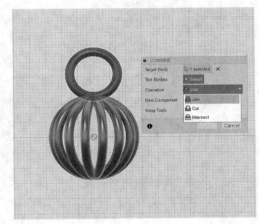

2. Click Join/Join.

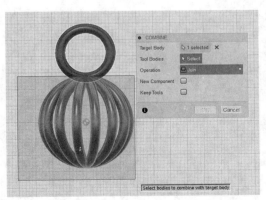

3. Select the cage.

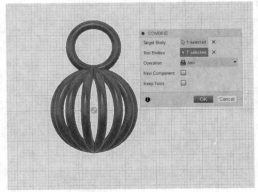

4. Click ENTER to finish.

Figure 17-12 Join all the parts together.

Now right-click on the model's name in the Browser window, in this case, "torus v4." Choose Save as STL (Figure 17-13). Accept the defaults in the dialog box that appears, and click OK.

Print It!

Figure 17-14 shows the model's orientation on the Simplify3D build plate and the supports generated by clicking Tools/Customize Support Structures (Figure 17-15). Double-click on the model to bring up dimension text fields. You can

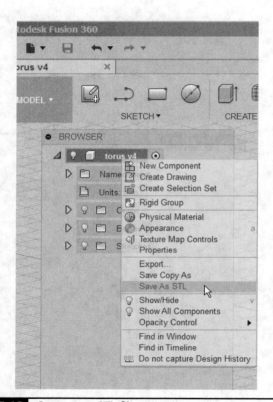

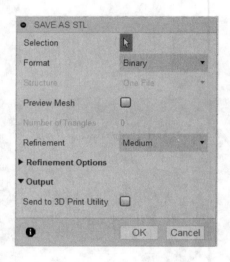

Figure 17-13 Save as an STL file.

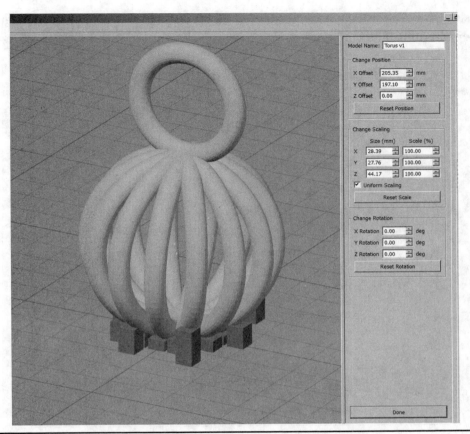

Figure 17-14 The model inside Simplify3D. The blocks at the bottom are supports.

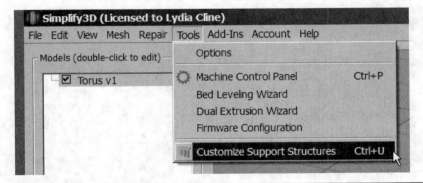

Figure 17-15 Generate supports.

accept the existing dimensions or type new ones. I chose a 40 percent infill. The temperature was 200°C, and rafts were enabled. Figures 17-16, 17-17, and 17-18 show other settings.

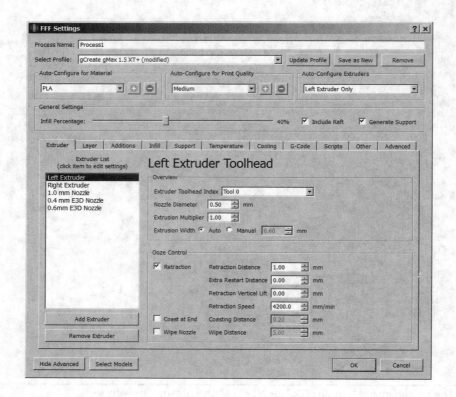

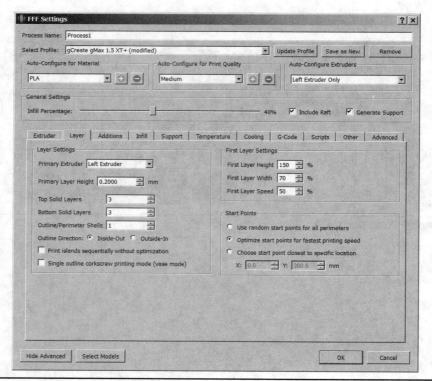

Figure 17-16 Extruder and layer settings.

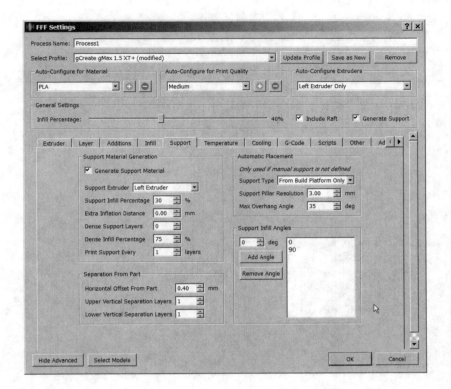

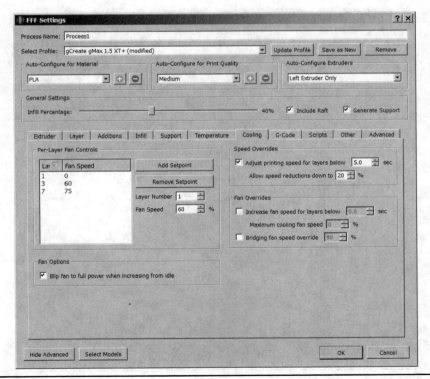

Figure 17-17 Support and cooling settings.

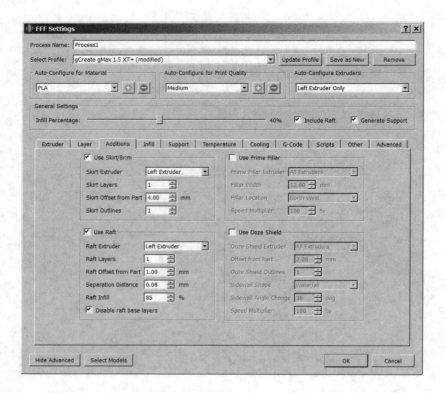

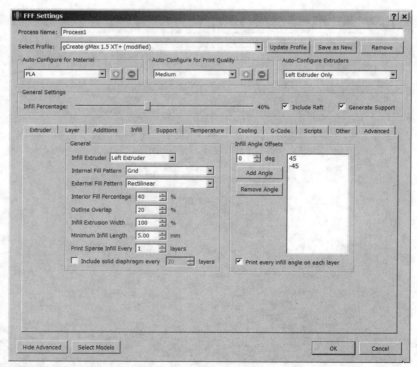

Figure 17-18 Additions and infill settings.

The pendant was printed on a cold aluminum build plate covered with BuildTak over which painter's tape wiped with isopropyl alcohol was placed. When the cage was developed enough to hold a small bead, I dropped one in (Figure 17-19). If you wait until the print is further along, it will hold a larger bead. Make sure that any item you add is not in the extruder path, and add it when the extruder is far from the area you're dropping it into. Figure 17-20 shows the finished print and its supports. Bronzefill is PLA infused with bronze powder, and needs to be post-processed (e.g., tumbled or wire-brushed) to look like metal.

Figure 17-19 A bead is added while the print is under construction.

Figure 17-20 The printed cage pendant and its supports.

Military Insignia Soap Mold

PROBLEM: A military spouse club wants to make soap featuring the unit's crest to sell as a fund-raiser. We'll use Tinkercad and Inkscape to digitally model a soap bar, print it on a Lulzbot Taz 6, and make a silicone mold from which a real soap bar can be cast.

Point your browser to tinkercad.com and log in to your Autodesk account. Tinkercad is a Web app. It saves your work automatically and saves it in the cloud. Once logged in, you'll see your dashboard. Click on *Create New Design* (Figure 18-1), and the Tinkercad workspace will appear.

Things You'll Need

Description	Source	Cost
Computer and Internet access	Your own or one at a makerspace	Variable
Autodesk account	autodesk.com/	Free
Inkscape	Inkscape.org	Free
Tinkercad Web app	Tinkercad.com	Free
3D printer and slicing software	Your own, one at a public makerspace, or one at an online service bureau	Variable
Thumb drive (needed only for offsite printing)	Computer or electronics store	< $10
Spool of PLA filament	Amazon, Microcenter, or online vendor	Variable
Oomoo 30 liquid rubber	amazon.com/Smooth-Silicone-Making-Rubber-30/dp/B004BNF3TK	$25 for a 2.8-lb package
Glue gun and glue stick	Amazon or discount store	< $15
Can of WD-40	Amazon or discount store	< $15
Three plastic cups, plastic container, stirrers	Amazon or discount store	< $5

Create New
Design

Figure 18-1   My Tinkercad dashboard. Click on Create new design.

Model the Base

On the right side of the screen is the Basic Shapes menu. Click on the box, drag it into the work plane, and then release the mouse (Figure 18-2).

When you release the box, grips for resizing and an Options window will appear. Drag the

black center/side grips to change the shape, and drag the white button on top to change the height (Figure 18-3). You can eyeball the changes or type exact dimensions into the text fields. Hold the SHIFT key down to scale it proportionately.

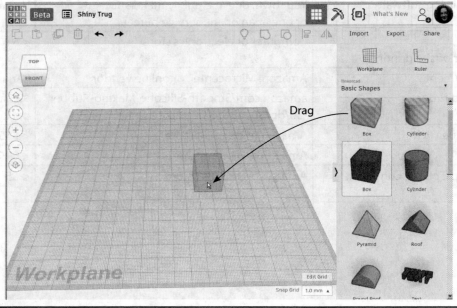

Figure 18-2   Drag a box into the workspace.

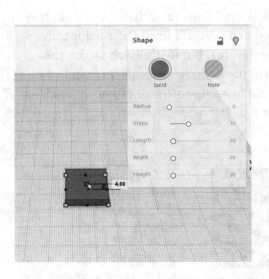
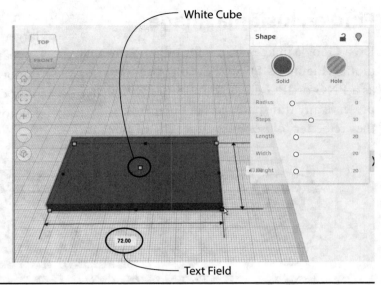

Figure 18-3 Change the shape of the box with grips.

The Options box has a Radius slider for smoothing the edges. While this makes a more interesting shape (Figure 18-4), those edges may need supports to print. So we'll keep the sharp edges.

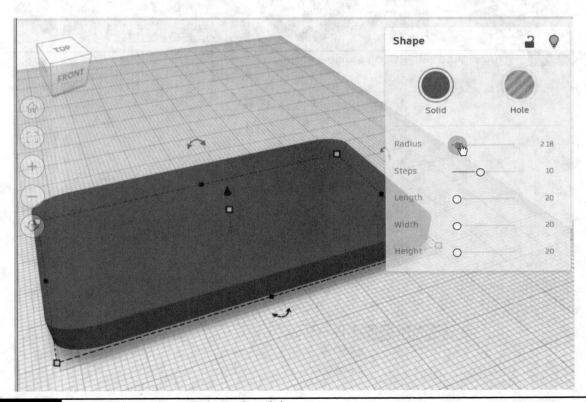

Figure 18-4 Smooth the edges with the Radius slider.

Find an Insignia Image

Do a Google images search for military insignia motifs (Figure 18-5). Save the one you want as a PNG or JPG file (Figure 18-6). PNG files preserve any transparencies in the image, so they're better to use. Files larger than 1,000 mm³

won't insert into Tinkercad. If the file is slightly larger, you can adjust it inside text fields that appear on import, but if it's much larger, the import will just fail. So reduce a large image in your digital imaging software of choice, such as Photoshop or Irfanview, first.

Figure 18-5   Pictures found with a Google images search.

Figure 18-6   The insignia image I chose.

Convert the Insignia Image to an SVG File

Launch Inkscape. Click File/Import, and bring in the insignia file, accepting the defaults in the dialog box (Figure 18-7). Move the file into the page outline that appears; size the file to fit inside the page by clicking on and dragging its grips. Hold the CONTROL key (OPTION key on the Mac) down to size the file proportionately.

When the file is selected (grips will appear); click on Path/Trace Bitmap (Figure 18-8). Accept the defaults in the dialog box that appears, and click OK to execute the trace

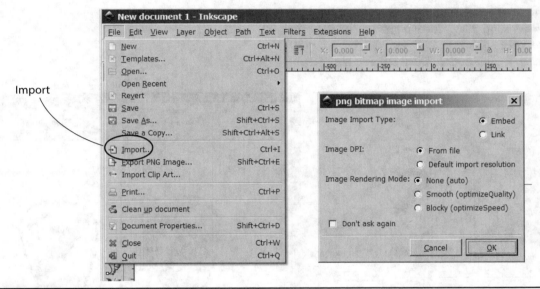

Figure 18-7 Import the insignia file into Inkscape, accepting the defaults in the dialog box.

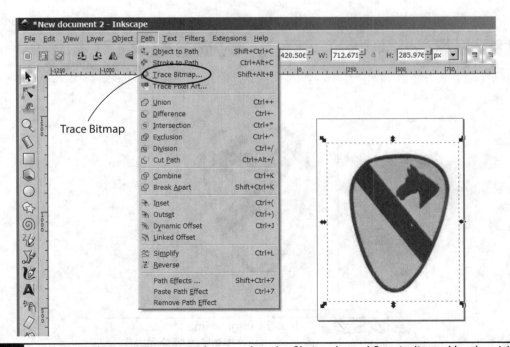

Figure 18-8 Click on Path/Trace Bitmap. Make sure that the file is selected first, indicated by the visible grips.

(Figure 18-9). Click the Update button to see the converted file in the Preview window, and then close the dialog box. The converted file will be on top of the imported image; click it and drag it off to the side (Figure 18-10).

> *Tip:* The website http://www.online-convert.com/ does the same thing as Inkscape's bitmap tracer. You may get better results with one than the other, depending on the file. The online converter did a poor job converting this particular image, but Inkscape did a good job.

Now delete the original image file, and drag the converted file back into its place (Figure 18-11). Then click File/Save as, click the dropdown arrow next to the Save as type field, and save it as a Plain SVG (Figure 18-12).

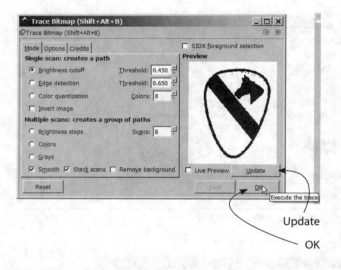

Figure 18-9 Click OK to trace and Update to view the new file.

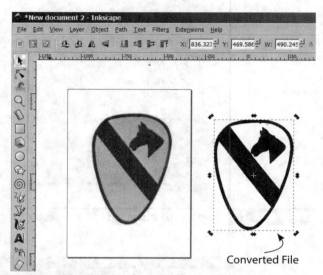

Figure 18-10 Drag the converted file off the original image file.

Figure 18-11 Delete the original image file, and drag the converted file back into its place.

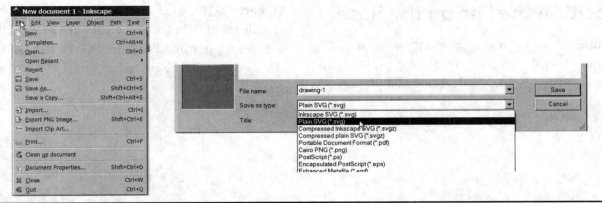

Figure 18-12   Click File/Save, and save as a Plain SVG.

Import the SVG File into Tinkercad

Return to the Tinkercad file. Click on Import, and bring in the SVG file. A dialog box appears that shows the size of the imported file (Figure 18-13). The file will be very large. You can type different dimensions in the dialog box's text fields or resize the file after import by dragging its center/side grips (Figure 18-14).

Figure 18-13   Import the SVG file into Tinkercad. A dialog box shows the imported file size; you don't type new dimensions in the dialog box's text field.

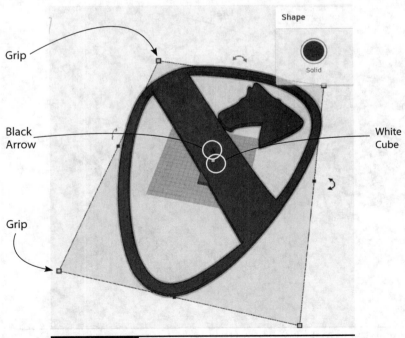

Figure 18-14   Drag the grips to resize.

Position the File on the Base

Rotate the file by dragging the curved arrow (Figure 18-15). Drag a work plane onto the soap bar to make positioning the imported file easier (Figure 18-16). Click the black arrow to move the file up and down, and click anywhere on the file to drag it around (Figure 18-17). Rotate the file by dragging the curved arrows.

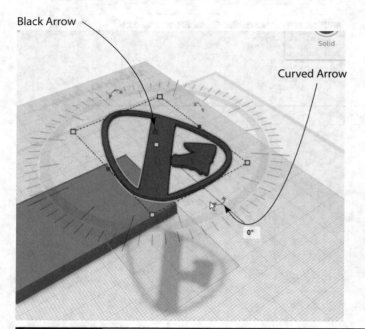

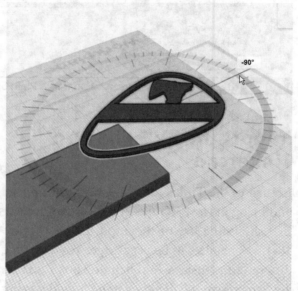

Figure 18-15   Rotate the file by dragging the curved arrow.

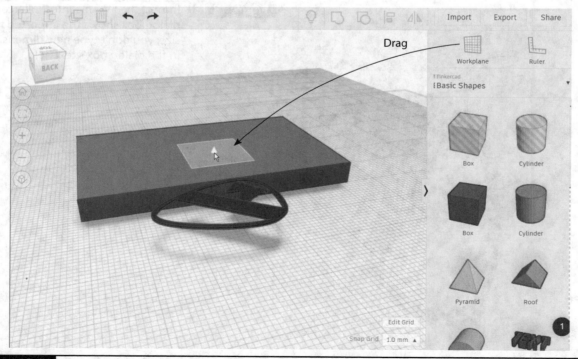

Figure 18-16   Drag a work plane onto the soap bar to make positioning the file on it easier.

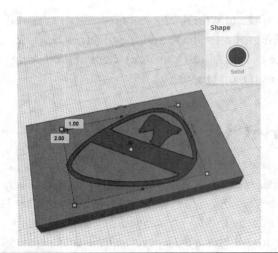

Adjust the height of the letters by dragging the white box up, and then push the letters down a bit into the base (Figure 18-18).

Figure 18-17 Click the black arrow to move the imported file up and down. Click anywhere on the file to drag it around.

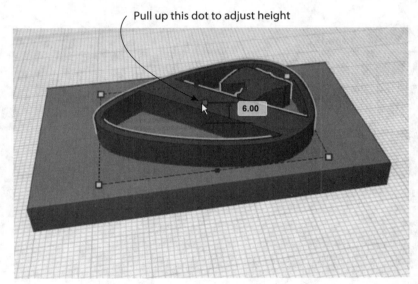

Pull up this dot to adjust height

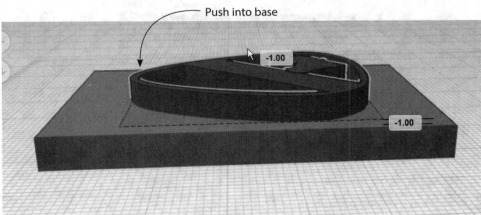

Push into base

Figure 18-18 Adjust the height of the letters as needed, and then push them down a bit into the base.

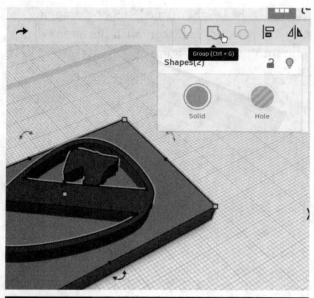

Figure 18-19 *Group the letters and base together.*

Make any final adjustments to the base and then group the base and file together by dragging a selection window around them and clicking the Group icon (Figure 18-19).

Print It!

Click on the Export button. A dialog box will appear; click on *Everything in the design*. Then save as an STL file (Figure 18-20). Figure 18-21 shows the model positioned in Cura and some settings. I chose a 15 percent infill and printed it on a heated borosilicate plate covered with PEI.

1.

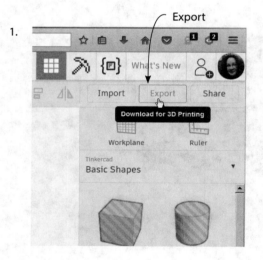

2.

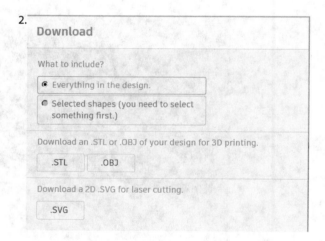

3.

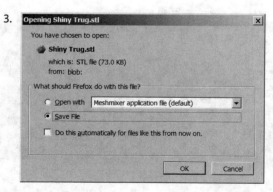

Figure 18-20 *Download the soap bar as an STL file.*

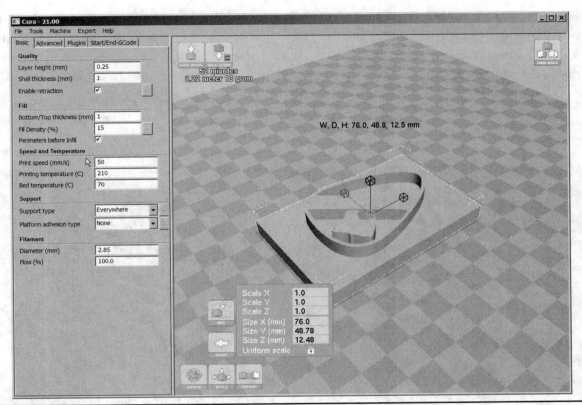

Figure 18-21 The model on the Cura build plate.

Make a Silicone Mold

A vacuum former is the best way to make a mold, but a cheaper option is silicone rubber. We're going to use a product by Smooth-On called Oomoo 30. Find a cuttable paper or plastic container that's a little bigger than the print. Small convenience food packaging works well. Then gather up stirrers, a can of WD-40, glue gun, plastic cups, and a box cutter (Figure 18-22).

1. Deposit a bead of glue around the print's bottom perimeter to attach it to the container (Figure 18-23).

Figure 18-22 The print and supplies for the silicone mold.

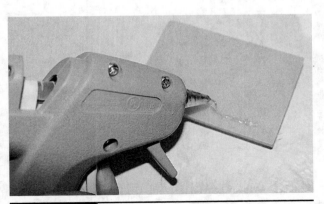

Figure 18-23 Glue the bottom of the print.

2. Spray the print with WD-40 (Figure 18-24) or smear it with another lubricant such as Vaseline. Empty out any excess that pools inside the container.

3. Stir the contents of bottles A and B, and then measure equal amounts of each into separate cups. Clear cups with measuring lines on them make this easier (Figure 18-25).

4. Mix the liquids together, and stir well for 3 minutes (Figure 18-26).

Figure 18-24 Spray or smear lubricant on the print.

Figure 18-25 Stir and measure equal amounts of the liquid in bottles A and B. Plastic cutlery makes good stirrers.

Figure 18-26 Mix the liquids and stir well.

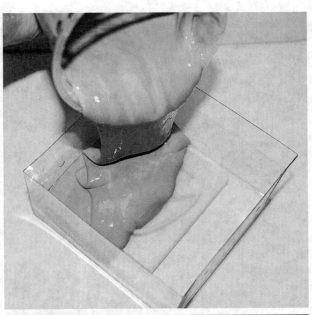

Figure 18-27 Pour the liquid over the print, and wait 6 hours.

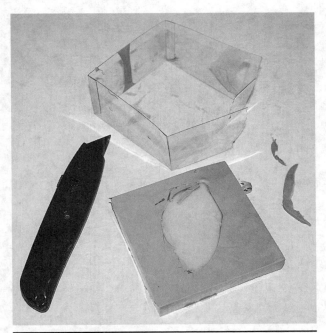

Figure 18-28 Cut the container off and any excess rubber as needed.

5. Pour the liquid completely over the print and wait 6 hours (Figure 18-27).

6. Cut the container off, and cut excess rubber from around the print as needed to remove it (Figure 18-28). The rubber mold can now be cast with soap or other liquid (Figure 18-29). This particular product is not food safe.

Further Resources

- Download Irfanview, a free digital imaging program: irfanview.com/main_download_engl.htm.

- Video that shows how to use Oomoo: www.youtube.com/watch?v=TgLKwMV7NbM.

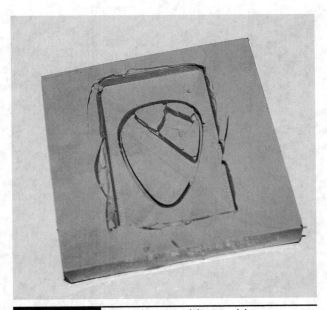

Figure 18-29 The silicone rubber mold.

Hanging Lampshade

PROBLEM: A young professional wants to upgrade the shade of a hanging light fixture bought while in college. We'll model one in Fusion 360, apply a Voronoi (open mesh) pattern in Meshmixer, slice it in Simplify3D, and print it in ABS on a Lulzbot Taz 6.

Figure 19-1 shows the light fixture, and Figure 19-2 shows rough dimensions. Figure 19-3 shows the interior diameter of one of the collars between which the new shade will fit. We'll make the diameter 2 mm smaller to fit. It will have a straight part to fit over the top of the bulb and a flared part to fit over the rest of the bulb. So launch Fusion 360.

Things You'll Need

Description	Source	Cost
Computer and Internet access	Your own or one at a makerspace	Variable
Hanging light fixture	Amazon, Ikea, or discount store	< $10
Autodesk account	autodesk.com/	Free
Autodesk Fusion 360 software	autodesk.com/products/fusion-360/overview	Free or subscription cost
Meshmixer software	meshmixer.com/	Free
3D printer and slicing software	Your own, one at a public makerspace, or one at an online service bureau	Variable
Thumb drive (needed only for offsite printing)	Computer or electronics store	< $10
Spool of ABS filament	Amazon, Microcenter, or online vendor	Variable

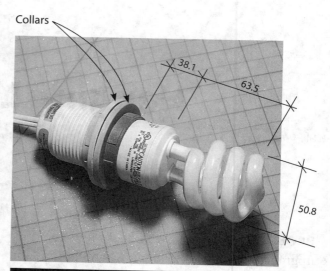

Collars

Figure 19-2 Dimensions in millimeters. The new shade will fit between the two collars just like the old shade.

Figure 19-1 The light fixture and its original shade.

Figure 19-3 Use a digital caliper to get the collar interior's precise diameter.

Model the Straight Part

On the Sketch menu at the top of the page, click the Circle tool (Figure 19-4). Then click on the horizontal plane, click on the work planes' origin, and type the circle's diameter, in this case 53.36 mm (Figure 19-5).

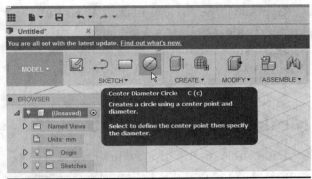

Figure 19-4   Click on the Circle tool.

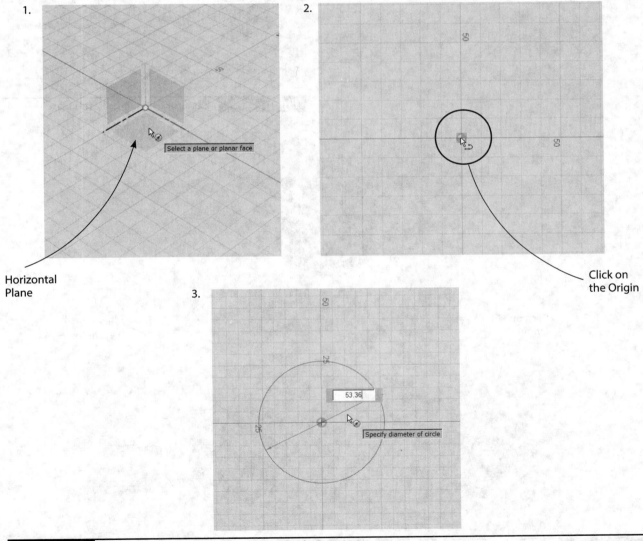

Figure 19-5   Sketch a circle, placing its center at the origin.

Right-click the circle, and click Press Pull. Then extrude it up into a cylinder that's the height of the bulb top, in this case 38 mm (Figure 19-6).

Model the Flared Part

Sketch a circle on top of a cylinder that's the same size as the first one (Figure 19-7). Then right-click on it, choose Press Pull, and extrude it up to cover the rest of the light bulb, in this case 70 mm. Drag the button on the handle to flare the cylinder 10 degrees (Figure 19-8).

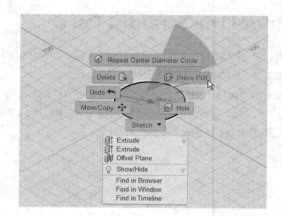

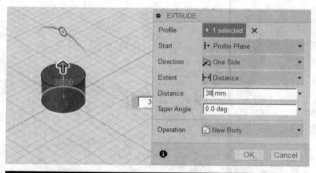

Figure 19-6   Extrude the circle sketch up.

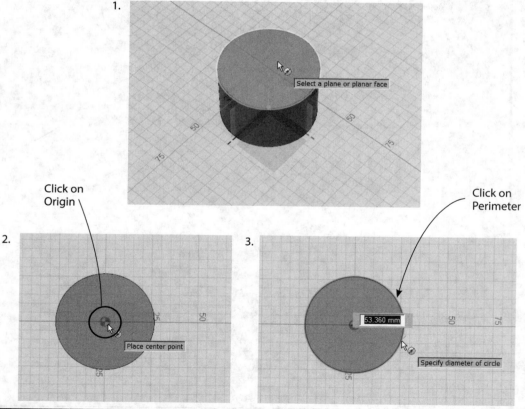

Figure 19-7   Sketch a circle on top of the cylinder

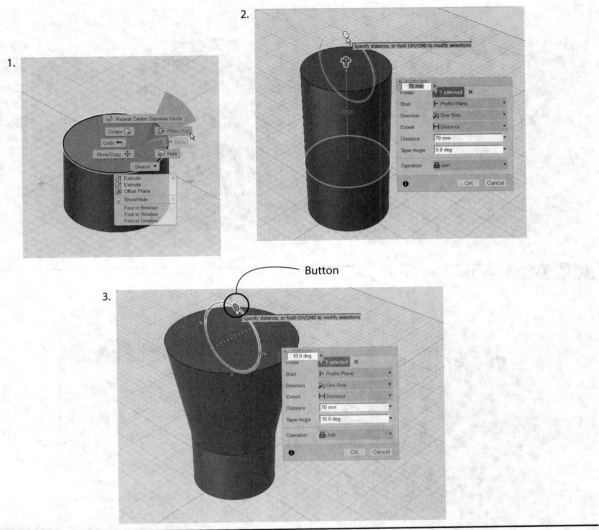

Figure 19-8   Extrude and flare a cylinder to cover the light bulb.

Shell and Fillet the Lampshade

Right-click on the flared face and click on Shell. Then drag the arrow an appropriate thickness (Figure 19-9) or type a dimension. Keep in mind that the thickness must enable the lampshade to fit inside the inner ring of the lamp holder; 3 mm (Figure 19-9) will work on this project. After shelling the lampshade will hollow out, but the narrow end will remain solid (Figure 19-10).

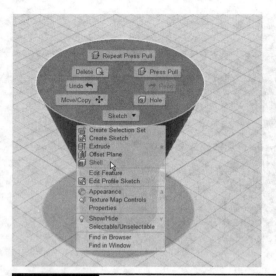

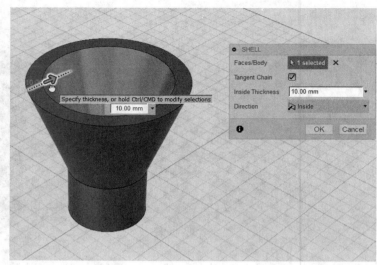

Figure 19-9 Shell the lampshade.

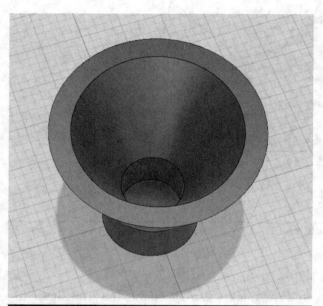

Figure 19-10 The narrow end remains solid after shelling.

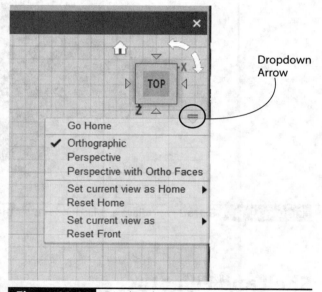

Figure 19-11 Display the workspace orthographically to make sketching easier.

To open the narrow end, we'll sketch a circle on it and extrude. Sketching is easier if we display the workspace orthographically. Hover the mouse over the View Cube in the upper-right screen, click the dropdown arrow that appears, and click Orthographic (Figure 19-11).

Click on the Circle tool. Click it on the center of the narrow end (again, which is the work plane's origin), and then click on the perimeter (Figure 9-12). Then right-click on the sketch, click on Press Pull, and extrude the circle down. Click to finish. That end will now have a hole (Figure 9-13).

Click on Origin

Click on Perimeter

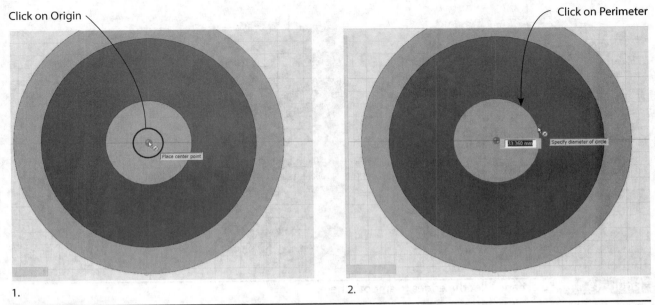

1.   2.

Figure 19-12 Sketch a circle onto the cylinder, giving it the same center and diameter.

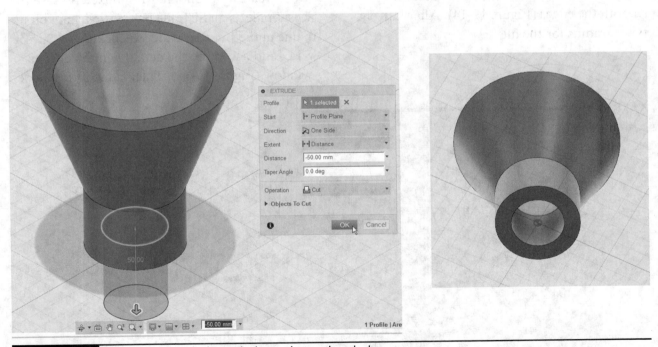

Figure 19-13 Extrude a sketch through the end to make a hole.

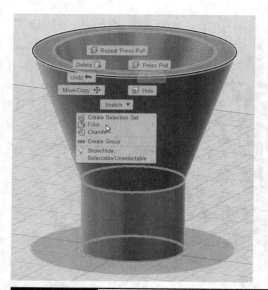
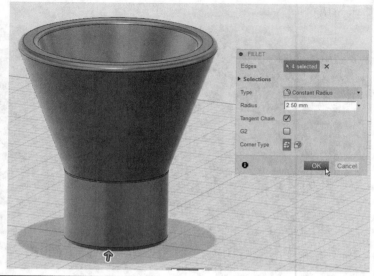

Figure 19-14   Select and fillet the edges to smooth them.

Smooth the edges by clicking on each to select them (hold the SHIFT key down to make multiple selections). Then right-click on any selected edge, click on Fillet, and drag the arrow to smooth the edges (Figure 19-14). Alternatively, type a radius for the fillet.

Save as an STL File and Import into Meshmixer

We'll create a Voronoi in Meshmixer. So save the Fusion file as an STL file by right-clicking on its name in the Browser window, clicking on Save as STL, and accepting the defaults in the options window (Figure 19-15). Then launch Meshmixer,

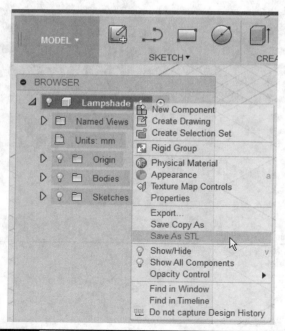
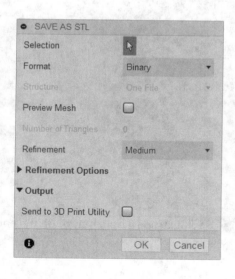

Figure 19-15   Save the lampshade as an STL file.

click on Import, and bring the STL file in (Figure 19-16). If it enters in an odd orientation, hit the T key, which brings up a manipulator. Drag the arrows to move the lampshade up, down, and sidewise, and drag the curves to change the angle. Drag the mouse directly over the radial marks to change the angle in whole degrees (Figure 19-17).

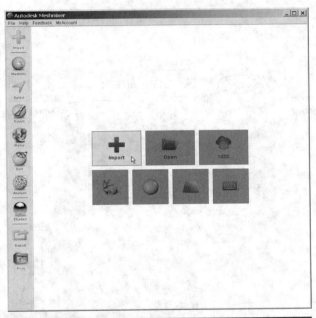

Figure 19-16 Click Import to bring the STL file into Meshmixer.

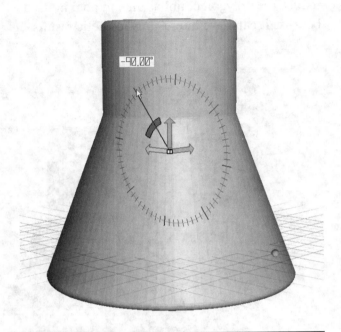

Figure 19-17 Hit the T key to bring up a manipulator for changing the location and orientation. Here the lampshade has been rotated 90 degrees.

Decimate the File

We can apply the Voronoi pattern now, but because of the file's high polygon count, the mesh holes will be small. A lower polygon count will produce bigger holes. As an aside, you can reduce the polygon count as an end goal in itself because it can have interesting aesthetic results.

Click the Select icon, click once on the lampshade, and then, under Select/Modify, click on Select All (Figure 19-18). The entire lampshade will select. Under Select/Edit, click on Reduce, and move the Percentage slider to the right (Figure 19-19). The more you move it, the more the lampshade's form will change.

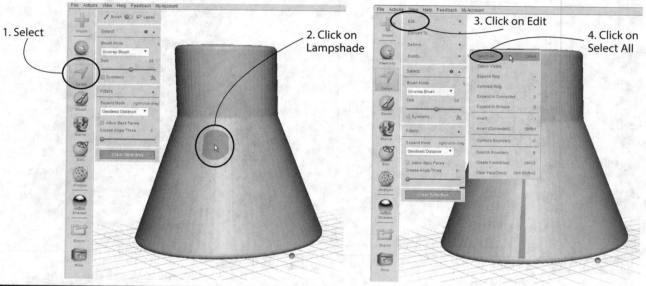

Figure 19-18   Select the lampshade.

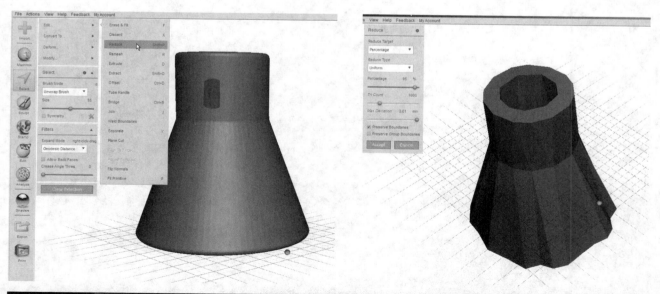

Figure 19-19   Reduce the lampshade with the Edit/Reduce option.

Apply a Voronoi Pattern

Click on the Edit icon, then click on Make Pattern. An Options box will appear; click in the Pattern Type text field, and click on Dual Edges (Figure 19-20). The Voronoi pattern will appear on the lampshade; click Accept.

Troubleshooting

If your lampshade appears muted after the dual-edges operation is finished, turn off one of the entries in the Object Browser box by clicking on the eye (Figure 20-21). If you don't see an Object Browser box, click on the View menu at the top of the screen and then on Show Object Browser.

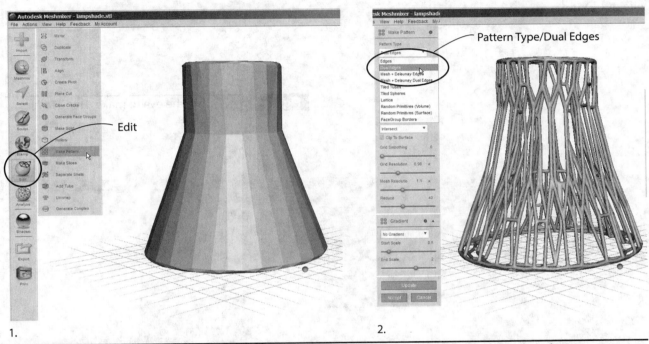

1.   2.

Figure 19-20   Apply a Voronoi pattern with Edit/Make Pattern/Dual Edges. This file was reduced 95 percent.

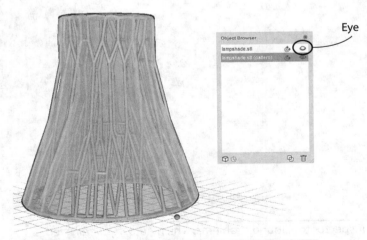

Figure 19-21   Turn off an entry in the Object Browser box to fix this muted appearance.

If your lampshade looks like Figure 19-22, you inadvertently hid it; click the eye next to its Object Browser box entry to turn it back on.

The Voronoi pattern may create holes or other unprintable problems in the file. Click on the Inspector icon, and then click on Analysis. Pins appeared, indicating problems (Figure 19-23). Click on *Auto Repair All* to fix. They were fixed with minimal change to the lampshade, so I clicked Done.

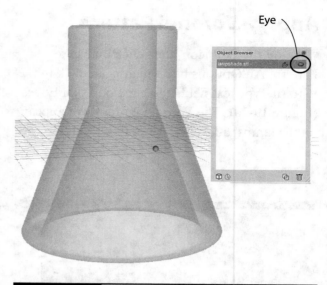

Figure 19-22 Click the eye to make a hidden file visible.

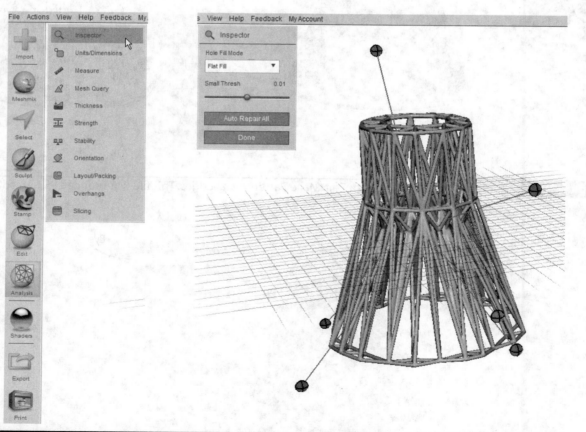

Figure 19-23 The Inspector tool found problems, which *Auto Repair All* fixed.

Iterate and Print It!

Figure 19-24 shows the model's orientation on the Simplify3D build plate. Supports were generated by clicking on Tools/Customize Support Structure. I used a raft, 15 percent infill, and ABS filament to withstand the light bulb's heat and printed it on a heated borosilicate bed covered with PEI wiped with isopropyl alcohol.

The print failed at about 40 percent done because it was too fragile for this printer. When prints fail, try changing the settings, nozzle size, or design. In this case, changing the settings

didn't help, so I imported the original lampshade into Fusion 360 again and reduced it less to make the holes smaller. I also adjusted other Dual Patterns options, such as making the individual elements larger. But all subsequent prints still failed, leading me to conclude that I either needed a different nozzle size, a machine more capable of such a fragile print (e.g., a resin or commercial printer), or a different pattern. I ended up printing the original Fusion 360 lampshade with no pattern, using a 5 percent infill and no raft or supports. Figure 19-25 shows it on the build plate, Figures 19-26 and 19-27 show some settings, and Figure 19-28 shows the final print.

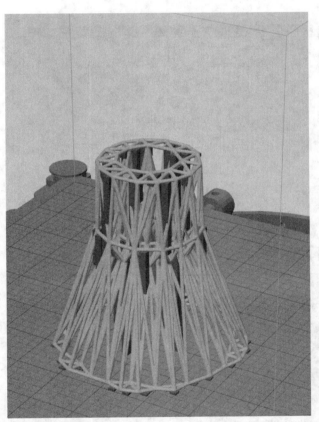

Figure 19-24 The model's orientation on the Simplify3D build plate.

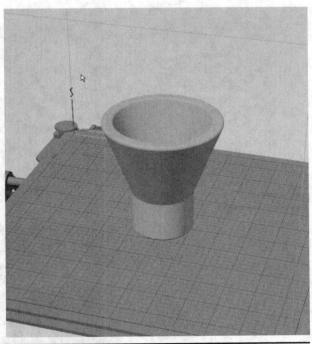

Figure 19-25 The final print on the Simplify3D build plate.

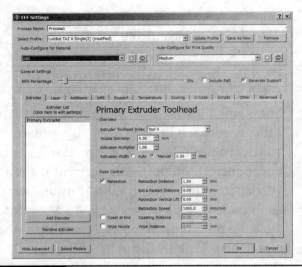

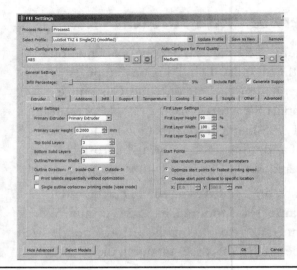

Figure 19-26 Extruder and layer settings.

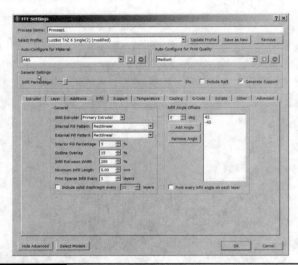

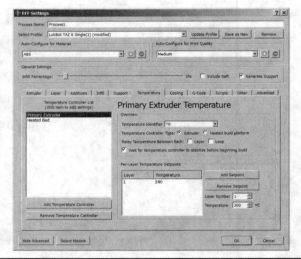

Figure 19-27 Infill and temperature settings.

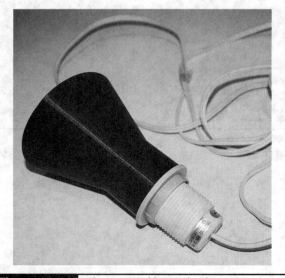

Figure 19-28 The printed lampshade.

Reality Capture
of a Buddha Charm

PROBLEM: A student wants to scan and print the family Buddha as a good luck charm to wear. In this project we'll use an SLR camera and Autodesk Remake to make a reality capture model of the statue in Figure 20-1. Then we'll add a loop to it in Autodesk Meshmixer, slice it in MakerBot Desktop, and print it on a MakerBot Rep 2.

What Is a Reality Capture?

A *reality capture* is a mesh model made from photos stitched together with local or cloud-based software. The technical name for this is *photogrammetry*, a technique that takes multiple overlapping photos of a subject and re-creates that subject by finding and matching the photos' common features.

Things You'll Need

Description	Source	Cost
Computer and Internet access	Your own or one at a makerspace	Variable
Autodesk account	autodesk.com/	Free
Autodesk Meshmixer software	Meshmixer.com	Free
Autodesk Remake software	remake.autodesk.com/try-remake	Free or subscription
3D printer and slicing software	Your own, one at a public makerspace, or one at an online service bureau	Variable
SLR (preferable) or cellphone camera	Online, warehouse or specialty shop	Variable
Thumb drive (needed only for offsite printing)	Computer or electronics store	< $10
Spool of PLA filament	Amazon, Microcenter, or online vendor	Variable

Figure 20-1   The Buddha statue we'll capture.

What Can Be Captured?

Reality capture is best done on subjects that have a matte, patterned, or textured surface. Bright colors work better than dark ones. Reflective, shiny, and transparent subjects—mirrors, glass, polished metal—don't give the processing software the information it needs. Ambiguous features such as hair, fur, and large swathes of plain fabric, or even symmetrical surfaces, are also poor subjects and often appear in the file as holes or bad mesh.

However, there are workarounds for capturing poor subjects. For instance, if a large part of the subject is plain, place colored sticky notes with black marker scribbles in the corners of that part. This will help the processor to find the common points it needs to correctly stitch the model together. Glass and other reflective subjects can be captured when covered with flour, talcum powder, or tempera paint (all these wash off easily).

How to Photograph a Subject

A successful capture requires methodical photographing. Here are some tips:

- *Place the subject where there is space to move completely around it.* Remove anything blocking your view.

- *Place the subject on an appropriate surface.* Choose one that isn't transparent or shiny and has a color and pattern that contrast with the subject. Gift wrap and newspaper work well (Figure 20-2).

- *Illuminate the subject with diffused light.* Consistent, even lighting all around the subject is critical. Photograph the subject in full shade, on an overcast day, or indoors under a fluorescent bulb. Avoid stark shadows, spotlights, strong backlighting, and direct sunlight on the subject because this results in under- or overexposed photos. Also avoid highlights because they move around the subject as you move around it.

- *All photos must be identically lit.* Don't use a flash because it illuminates each photo differently. Camera phones have an

Figure 20-2   The statue is on a kitchen island and on top of gift wrap.

automatic exposure feature, making their flash photos especially bad.

■ *Move 360 degrees sequentially around the subject* (Figure 20-3). Fill each frame with

the whole subject. Take shots every 20 degrees, maintaining the same distance from the subject. Shoot a second sequential circle at a different height. If needed, crouch down and shoot a third circle at a lower height. Then move in for any detail shots needed, such as of features blocked by other features or very complex features. Fill the frame with the whole subject on those shots, too. If the capture has more background than subject, you didn't shoot the subject closely enough.

■ *Overlap the photos.* Every feature should appear in at least three or four different photos taken at varying angles. Missed spots will appear as holes in the model.

■ *Upload the proper number of photos.* Between 40 and 80 photos gives best results for small objects (Figure 20-4).

■ *The subject must remain stationary.* Don't move or lift it to photograph hidden areas or underneath it. All features must be in the same place for good stitching.

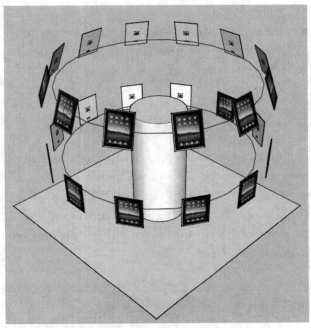

Figure 20-3 Move sequentially around the subject, taking photos at regular intervals.

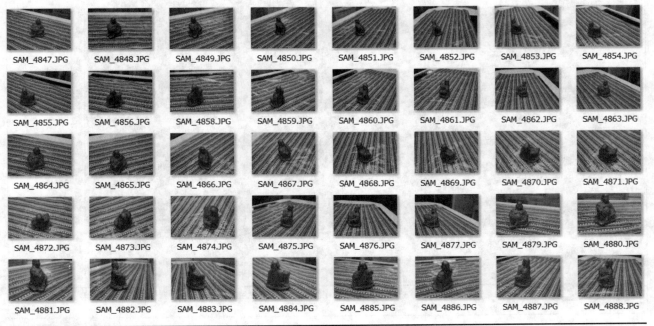

Figure 20-4 The photo set for the Buddha statue.

- *Quality of the camera.* Good cameras take better photos than bad ones, especially in low lighting. The results show in the capture.

- *Take raw images.* Turn off camera features such as sharpness enhancement, image stabilization, and anything else that artificially modifies the photo. Such "interpreted" photos don't process as well.

- *Focus.* Images must be sharp because blurry ones don't provide the feature information needed. Hold the camera against your body, or put it on a tripod for extra stabilization. If you change the aperture setting to blur the background via depth of field, do that in all the photos. An 18-mm focal length is a good setting.

- *Resolution.* More resolution equals more detail in the model.

- *Orient all photos similarly.* All photos need to be either portrait or landscape.

- *Upload the photos as is.* Don't crop or edit. You can downsize their resolution to speed up processing, but that will affect the model's detail and quality.

- *Delete bad photos before uploading.* Blurry, overexposed, or underexposed photos mess up the stitching process.

Photographing a Person

Get best results when photographing a person by seating him or her and telling him or her to avoid small facial movements. The subject should look forward, blink between shots, and close his or her mouth. If the subject moves even a little, start over.

Tip: If you want to model a celebrity, find a bobble-head statue of that person, and take photos of that (Figure 20-5).

Figure 20-5　To capture a celebrity, take photos of a bobble-head statue.

Upload the Photo Set to Remake

Launch Remake and click on the Create 3D button (Figure 20-6). A dialog box appears asking you to create the model offline or online. The free version of Remake only allows for online processing, so I clicked that. Another dialog box appears, asking where the photos are—on your computer or in your Autodesk A360 account. I clicked Local Drive (Figure 20-7), navigated to the photo set, and hit ENTER to upload it.

> **Tip:** A360 is an Autodesk site that offers storage and collaboration ability. Make a free account at a360.autodesk.com/drive.

Create 3D

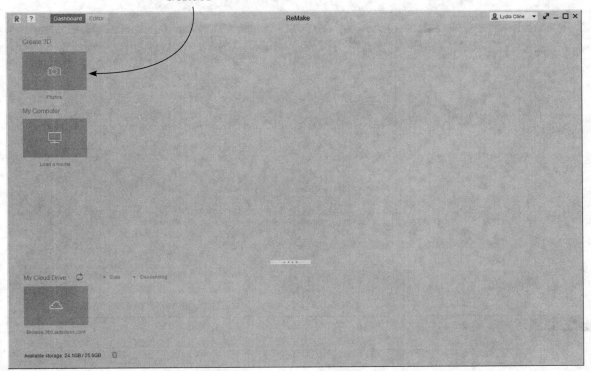

Figure 20-6 Click on the *Create 3D* button.

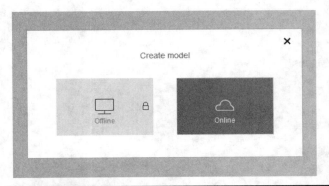
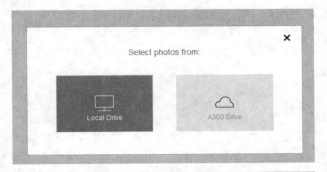

Figure 20-7 Click *Online* for processing and *Local* for the photo set location.

Create the Model

Once the photo set is uploaded, click *Create model* (Figure 20-8). A dialog box appears (Figure 20-9). Name the file and keep the defaults.

The model will begin processing; a small item typically takes 15 to 20 minutes to complete. A thumbnail view will appear while processing. When complete, a Ready to Download notice appears (Figure 20-10). Hover your mouse over the thumbnail to make a download arrow

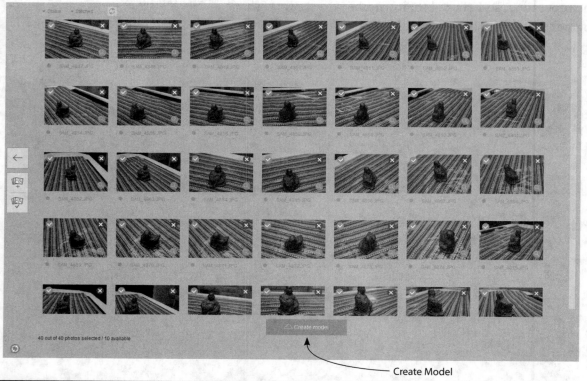

Create Model

Figure 20-8   Click *Create model* once the photo set is uploaded.

Figure 20-9   Name the model.

RCM Thumbnail

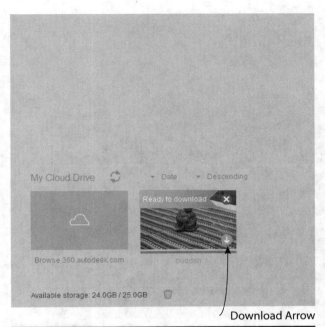

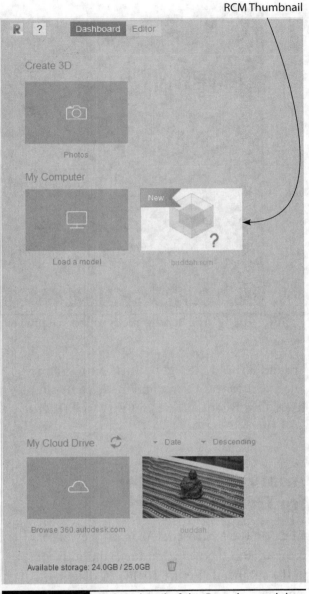

Download Arrow

Figure 20-10 Download the file from the thumbnail.

appear, and then save. A folder will appear that contains the RCM (Remake) file plus MTL, OBJ, and JPG files (Figure 20-11). The last three files are needed if you want to 3D-print the file's color and texture. Keep them all in the same folder.

Open the Model in Remake

After you save the RCM file, another thumbnail will appear (Figure 20-12). Click on it to open the model in Remake, where you can orbit

Figure 20-12 A thumbnail of the Remake model.

buddah.mti   buddha.jpg   buddha.obj   buddha.rcm

Figure 20-11 The Remake file folder and the four files inside it.

Figure 20-13   The Buddha model open in Remake.

around it (hold the right mouse button down), analyze it for problems, and make some simple fixes. This Buddha came out very well (Figure 20-13)!

Analyze the Model for Defects

Click on the Microscope icon on the vertical toolbar, and then click on the Band-aid icon in the submenu (Figure 20-14). A diagnostic window will appear. Click *Detect issues* and it will show what it finds (Figure 20-15). Here the analysis found a hole in the mesh and fixed it when I clicked the Fix button.

Figure 20-14   Click the Microscope icon to analyze the file.

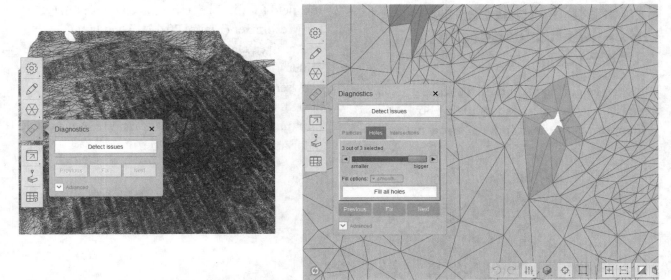

Figure 20-15   The diagnostics found a hole in the mesh and fixed it.

If you want to get rid of most of the background, drag the cursor around the part to keep, right-click, and click on Isolate Selection (Figure 20-16). Then hit the ENTER key. Everything outside the selection box will get deleted. However, I'm going to import the whole thing as is into Meshmixer and delete the background there.

Figure 20-16   Remove excess background by selecting, isolating, and deleting.

Export the Model as an OBJ File and Import into Meshmixer

Click the Export icon, and then click the *Export model* icon in the submenu. A dialog box will appear; click the dropdown arrow in the text field for file format options. I chose OBJ for Meshmixer (Figure 20-17). Then launch Meshmixer, click Import, and navigate to the folder in which the OBJ file is located. The browser will only show the OBJ file (Figure 20-18), but Meshmixer reads the MTL and JPG files that are in the folder with it and imports their color and texture information.

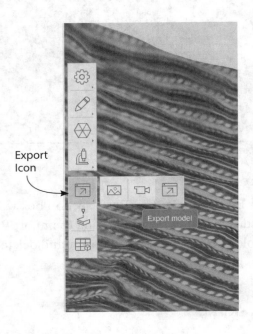

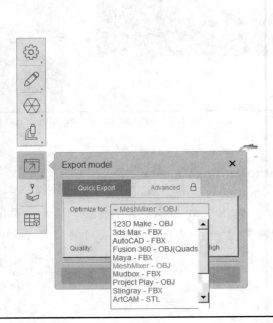

Figure 20-17   Export the RCM file as an OBJ file.

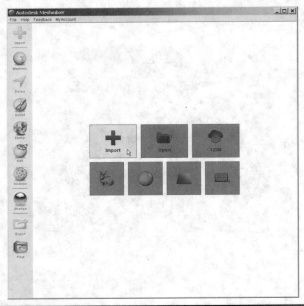

Figure 20-18   Only the OBJ file appears in the navigation browser, but the MTL and JPG files are being read.

Delete the Background

We don't need the color and texture information for this project, so click on the Shaders icon, and slide the gray shader over the model. This makes it easier to read (Figure 20-19). Next, click the Select icon, click on the background, and then drag the mouse all over the background (Figure 20-20). You can adjust the size of the brush with the Size slider. You can also draw a lasso with the mouse around bits and pieces that didn't get deleted. Once you've selected all the background, you can hit the DELETE key.

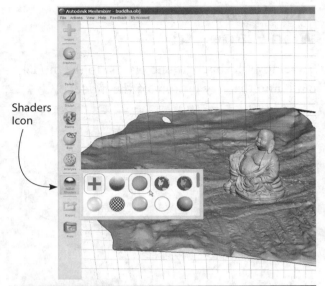

Shaders Icon

Figure 20-19 Slide the gray shader over the model to make it easier to read.

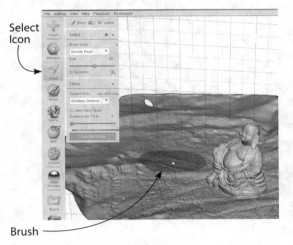

Select Icon

Brush

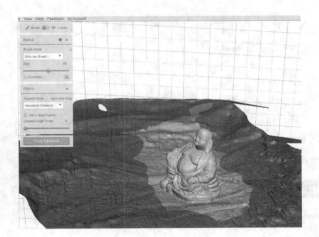

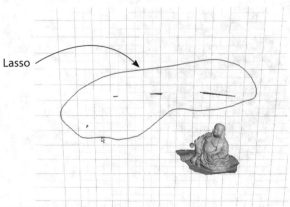

Lasso

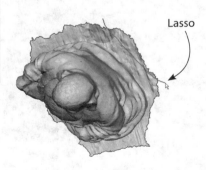

Lasso

Figure 20-20 Select and delete with the selection brush and lasso.

Cut the Bottom Off with Plane Cut

To remove the background closest to the model, cut it and the very bottom of the model off.

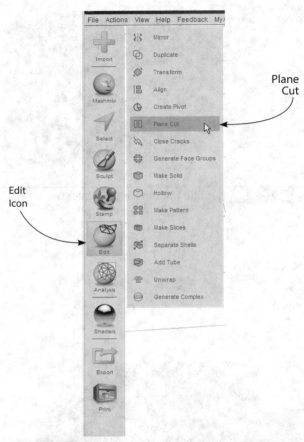

Edit Icon

Plane Cut

Figure 20-21 Click on Edit/Plane Cut.

Click on Edit/Plane Cut (Figure 20-21). The Transform tool will appear. Position the plane with the arrows and curves. The thick purple arrow points in the direction of what will be cut off, so make sure that it points down (Figure 20-22). Keep the defaults in the Plane Cut dialog box of *Cut (Discard Half)* and *Remeshed Fill*. The result should look like Figure 20-23.

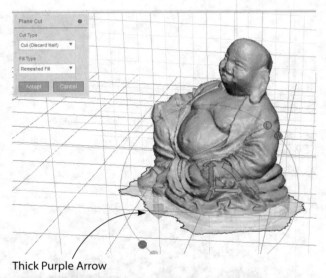

Thick Purple Arrow

Figure 20-22 Position the transformer widget at the bottom of the model.

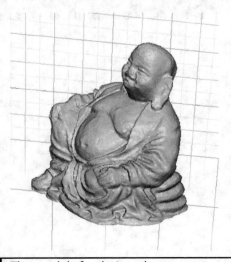

Figure 20-23 The model after being plane cut

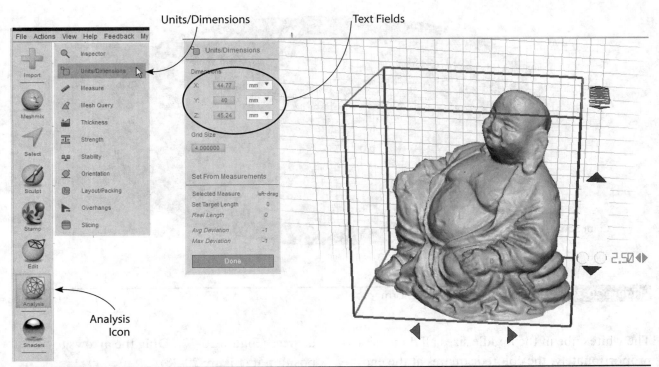

Figure 20-24   Click on Units/Dimensions, and type the desired size in a text field.

Resize the Model

Click on the Analysis icon, and then click on
Units/Dimensions. A box will surround the
the model with numbers showing its current
dimensions (Figure 20-24). Type a the size you
want in one of the boxes, and the rest will adjust
proportionately. I typed 40 mm in the *y*-field.

Add a Loop and
Combine All Parts

Click on the Meshmix icon, drag the torus into
the workspace—but not onto the model—and
let go (Figure 20-25). It will be very large; hit the
т key and size it down with the widget.

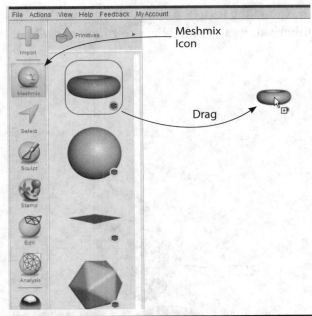

Figure 20-25   Drag a torus into the workspace.

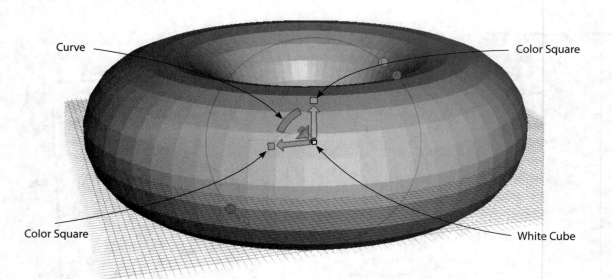

Curve

Color Square

Color Square

White Cube

Figure 20-26   Size it down with the Transformer tool.

The white cube in the middle sizes all dimensions proportionately; the colored squares at the end of each arrow size it along those dimensions (Figure 20-26).

Rotate the torus with the curves; drag the mouse along the radial lines to rotate it in whole

degrees (Figure 20-27). Drag the arrows to position it (Figure 20-28).

Finally, combine the torus and model by selecting both files in the Objects Browser

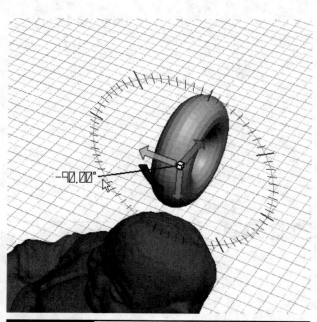

Figure 20-27   Rotate the torus with the Transformer tool curves.

Figure 20-28   Drag the arrows to position the torus.

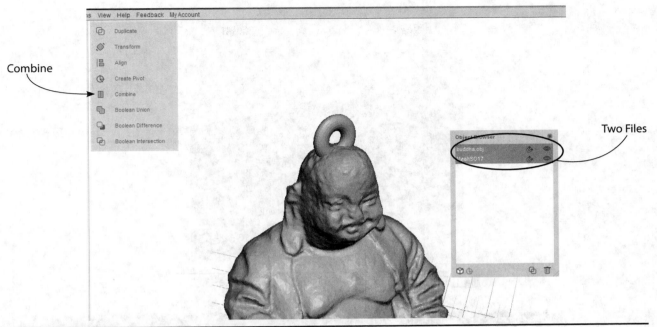

Figure 20-29 Combine the torus and Buddha files.

box and clicking Combine in the dialog box that appears (Figure 20-29). If you can't see the Objects Browser box, click on View/Show Objects Browser at the top of the screen.

Print It!

Click the Export icon on the vertical menu, and then click on the dropdown arrow in the *Save as*

type text field to see the export options (Figure 20-30). I exported it as an STL file and printed it on a cold acrylic build plate with painter's tape wiped with isopropyl alcohol. Figure 20-31 shows the model's orientation on the build plate, and Figure 20-32 shows some settings. It has a 15 percent infill.

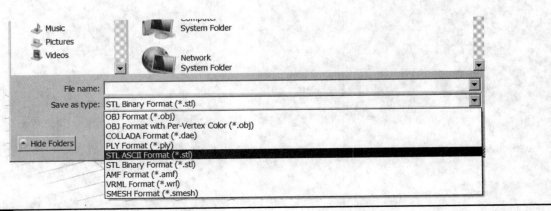

Figure 20-30 Export as an STL file.

Figure 20-31   Orientation on the build plate.

Figure 20-32   Settings for the model.

Figure 20-33 shows the finished print. Detail at this size (about 2" tall) is visible but would show up better with a commercial or resin printer.

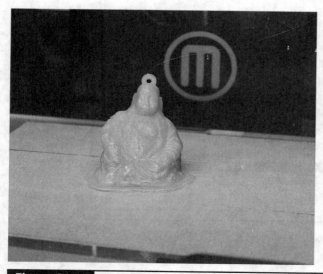

Figure 20-33   The Buddha print.

Post-Process All the Things

POST-PROCESSING IS THE APPLICATION of materials and techniques that enhance a 3D print's appearance. Prints can be filed, filled, drilled, sanded, smoothed, glued, stained, and painted (Figure PP-1). This chapter covers post-processing tools, techniques, and materials that are accessible to the average maker.

Sanding and Sandpaper

Sanding creates a smooth surface for painting. It also minimizes the seams where parts are joined together. Sandpaper comes in different grits and in block, stiff sheet, flexible sheet, and emery board form (Figure PP-2).

For ABS, 100 to 300 grit removes stringing and support remnants; 600 grit removes layer lines. ABS is easy to sand, so take care not to oversand because that will create depressions or remove critical dimensions. PLA needs a higher grit, such as 600 to remove stringing and supports and 1,200 to 2,000 grit for painting preparation. However, PLA becomes hot with sanding and can deform. If a PLA print needs a lot of work, it's better to cover it with primer and fillers and sand those materials instead.

Sand in small circular movements evenly across the print. Don't sand in one direction or sand parallel to the layer lines. You can sand dry or damp. Damp sanding reduces dust, clogging, and abrasiveness. Dip the print in water, and apply to primer. This technique can be more effective on multiple primer coats than dry

Figure PP-1   This wood print was sanded and stained.

Figure PP-2   Flexible and block sandpaper.

sanding. It's a separate sandpaper than what is used for dry sanding.

Files and Drills

Metal files smooth and remove stubborn bits of supports, rafts, and overextrusions and make small holes larger. If you are fitting two pieces together and one is too large, a file may often fix that. Files come in different sizes and shapes (Figure PP-3).

Drills do the same thing as files and sandpaper but with more firepower. The Dremel brand drill and its many accessories (Figure PP-4) are popular for removing stubborn rafts and supports, polishing rough edges, making small holes larger, and much more. Use a Dremel on low speed when drilling PLA, otherwise the filament will melt on the bit.

Figure PP-3 Files remove rough parts and make small holes larger.

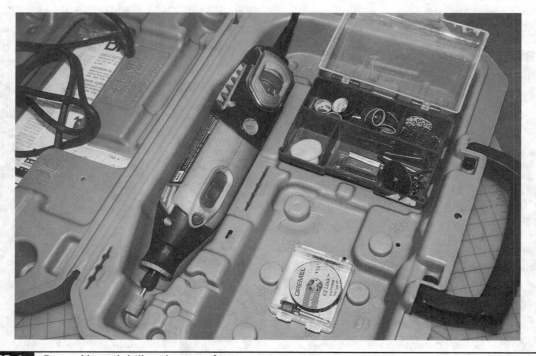

Figure PP-4 Dremel brand drill and some of its many accessories.

General Tools

Three tools needed when a print is finished are shown in Figure PP-5. They are

- *Flathead screwdriver.* This removes small prints from the build plate.

- *Scraper.* A rigid stainless steel one removes large prints from the build plate. This isn't the same as a spatula, which is a flexible tool.

- *Cutter.* This removes stubborn bits of rafts and supports (Figure PP-5).

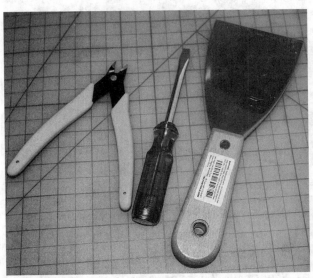

Figure PP-5 | Cutters, flathead screwdriver, and scraper.

Primers

Primers (Figure PP-6) fill holes, conceal imperfections, help paint to adhere to the print, and keep it from bleeding through the layers (filament is porous both on the spool and in the final print). Primers are also sandable. Automobile spray primer is high fill and works well on simple prints. Gesso art primer works well on detailed prints and comes in liquid and semisolid forms. Follow the manufacturer's instructions for drying time, and let it dry thoroughly before sanding or applying another coat. All primers require multiple, sanded coats to create a paint-ready surface.

Figure PP-6 | Primers prepare a surface for printing.

Paints

Paints are either acrylic (water based) or enamel (oil based). Acrylics are easier to work with, enable color layering, and wash off with water. Enamels require turpentine or paint thinner to remove and to clean the brushes. However, enamel adheres better and has richer, deeper tones. Paint comes in solid colors, glitter, glossy, matte, and textured. The model and craft paints sold at discount stores work well. Keep failed prints to practice on. A large cardboard box makes a good spray booth (Figure PP-7).

You can airbrush, spray, or apply paint with brushes, sponges, or pens (Figure PP-8).

- *Spray paint.* This type obscures a print's small details but works well on large, simple prints. Soak the can in warm water for a few minutes before using to get a thinner mist

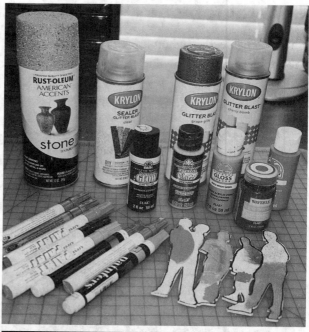

Figure PP-8   Spray, bottle, and pen paints in both acrylic and enamel. Use a failed print to experiment with colors.

and more even coating. Some textured paint, such as stone, is so thick that it doesn't need primer.

- *Bottled paint.* This is applied with brushes or sponges. It offers a lot of control and the ability to create fine details.

- *Paint pens.* These are easy to use and cover a lot of surface quickly.

- *Brushes (Figure PP-9).* These come in different sizes and tip shapes, and with synthetic or natural bristles. A selection of tip shapes enables you to achieve different effects and paint hard-to-reach areas and fine details. Sponges enable marbling and some primers are best applied with them. Experiment with different media (Figure PP-10).

Figure PP-7   Use a cardboard box as a spray booth.

Figure PP-9   Bristle and sponge brushes.

Figure PP-10   XTC-3D, spray paint, liquid paint, and markers were all used on this print.

- *Sharpie ink markers (Figure PP-11).* This common office supply item sticks to prints even without primer (Figure PP-12). These markers are available in multiple colors including gold, silver, and bronze.

Figure PP-11   Sharpie ink markers stick well directly to filament.

Figure PP-12   This print was sprayed with silver glitter and embellished with a silver Sharpie.

Other Finishes and Materials

- *Alcohol ink.* If you want some color but not the complete opacity of paint, soak an ordinary whiteboard marker in a small jar of 91 percent isopropyl alcohol, and then brush the colored mixture onto the print.

- *Clear finish.* If you don't want to color a print at all but want some kind of finish, apply layers of clear polyurethane, which is a liquid plastic available in spray or liquid form. Minwax clear gloss spray and Rustoleum Triple Thick Glaze spray work well. Sand each coat and apply a final top coat. Apply the liquid with sponges. Polyurethane doesn't withstand heat well, so this treatment is only for decorative pieces.

Other materials that come in handy include (Figure PP-13):

- *Loctite Gel Glue.* This holds print parts together.

- *Bondo.* This putty fills in holes, cracks, gaps, and covers seams. It can be sanded and painted. Large prints often have holes in them that need puttying (Figure PP-14). In fact, using an expensive filament on a large print may be a waste of money if the print has to be puttied and painted.

Figure PP-13   Brasso, Loctite Gel Glue, and Bondo.

- *Brasso.* This can be used as part of the post-processing of bronze filament (Figure PP-15).

Figure PP-14 Bondo was applied with a putty knife to the holes. Sand and paint.

Figure PP-15 This bronze filament print was sanded and rag wiped with Brasso. A few highlights were added with a bronze Sharpie marker.

Discount Store Crafts Section

Lots of things can be found in the crafts section of a hobby or discount store, such as art tape and wobbly eyes. Chartpak art tape comes in thin, black rolls and is available in multiple widths. It can substitute for small areas of black paint in places too hard to reach (Figure PP-16). Electrical tape, which comes in wider widths, also can be used.

Smoothing and Concealing Layers

The layers on both PLA and ABS can be smoothed (melted) or concealed to give the print a manufactured look. Do this in a well-ventilated area because the materials involved have strong and flammable fumes.

Figure PP-16 The black eyes on these prints are Chartpak art tape. The left print has wobbly eyes glued on with Loctite Gel Glue, the middle print's ears were colored with an enamel paint pen, and the right print was glitter spray painted.

Figure PP-17 This print was brushed with acetone for a rough, layerless appearance.

Figure PP-18 Supplies for an acetone vapor bath. A rice cooker works well for the larger container. Some nail polish remover is 100 percent acetone.

Smooth ABS a Little by Brushing Acetone on It

Lightly brush acetone onto an ABS print to melt it a little, creating a semismooth appearance. The print in Figure PP-17 was brushed with acetone just enough to remove the layers. Know that you can also brush the inside of an ABS container with acetone and the inside of a PLA container with epoxy to make them watertight (but not food safe).

Smooth ABS a Lot with a Cold Vapor Bath

Pour a capful of 100 percent pure acetone in a small container, and place it and the print inside a larger, covered container (Figure PP-18). Alternatively, soak paper towels in acetone and place them on all sides of the larger container. Don't put the print *in* the acetone or let the paper-soaked acetone touch the print. You could also put the print on a piece of foil or in a shallow container to enable removal without touching it.

Over a few hours the print will smooth (Figure PP-19). Check it frequently during this process because too much time in the bath will melt it into an undistinguishable mass or make it permanently tacky. Remove the print when it is smoothed to your liking, and let sit for a few minutes for the acetone to evaporate.

Figure PP-19 An ABS print before and after an acetone bath. (*Photographed by John Ewing for Andrew Sink, www.sinkhacks.com.*)

Chemicals for Smoothing PLA

Some filaments are a blend of PLA and PHA and will smooth somewhat in an acetone bath, but pure PLA does not react to acetone. Some makers have luck using tetrahydrofuran or ethyl acetate. Pour it onto a lint-free, non-dyed rag, and polish the print in circular motions. Let dry when finished.

Concealing Layers with XTC-3D

XTC-3D by Smooth-On is a thin, self-leveling epoxy that conceals a print's layers (Figure PP-20). It doesn't smooth the layers; it just covers them. This product comes in two bottles, and you mix their contents in unequal ratios. Clean and dry the print, apply the mixture with sponge brushes, and let dry overnight. Then prime and paint.

Concealing Layers with Putty and Primer

Both ABS and PLA can be covered with putty to fill in gaps. Cover the putty with several coats of high-fill primer (sand each coat separately after it dries), and top off with a thin primer. This treatment will make a print look smooth under paint, too.

Summary

Many treatments and techniques can be applied to a 3D print to make it look like an art piece or manufactured item. There are websites and YouTube channels devoted to painting and post-processing 3D prints. Find ones that interest you, practice different techniques, and take your prints beyond the build plate.

Figure PP-20 XTC-3D is an epoxy that coats a print so that layers won't show underneath paint.

Index

Page numbers in italics refer to figures.